# Solid Waste Management - Recent Advances, New Trends and Applications

*Edited by Pengzhong Li*

Published in London, United Kingdom

Solid Waste Management - Recent Advances, New Trends and Applications
http://dx.doi.org/10.5772/intechopen.1000359
Edited by Pengzhong Li

Contributors
Boniface Yeboah Antwi, Ammishaddai Ofori-Nyarko, Nazish Huma Khan, Nida Naz, Mohammad Nafees, Nida Gul, Tooba Saeed, Yasemin Tabak, Ayo Adeniran, Lorato Motsatsi, Sijekula Mbanga, Emma Ayesu-Koranteng, Winston Shakantu, Parin Somani, Atinuke Chineme, Marwa Shumo, Getachew Assefa, Irene Herremans, Barry Wylant, Lorenzo De Carlo, Giorgio Cassiani, Rita Deiana, Gian Piero Deidda, Maria Clementina Caputo, Iva Ćurić, Luka Brezinščak, Davor Dolar



First published in London, United Kingdom, 2023 by IntechOpen
IntechOpen is the global imprint of INTECHOPEN LIMITED, registered in England and Wales, registration number: 11086078, 5 Princes Gate Court, London, SW7 2QJ, United Kingdom

British Library Cataloguing-in-Publication Data
A catalogue record for this book is available from the British Library

Additional hard and PDF copies can be obtained from orders@intechopen.com

Solid Waste Management - Recent Advances, New Trends and Applications
Edited by Pengzhong Li
p. cm.
Print ISBN 978-1-83769-811-0
Online ISBN 978-1-83769-810-3
eBook (PDF) ISBN 978-1-83769-812-7

# Meet the editor

Dr. Pengzhong Li is a professor at the School of Mechanical Engineering, Tongji University, Shanghai, China. He received his Ph.D. in Mechanical Engineering from the same university in 2004. From 1995 to 2001, he was a manager in the Business Department and Warehousing Management Department of Guilin Daewoo Bus Co., Ltd. He is the Director of the Intelligent Manufacturing and Services Branch of the China Creative Studies Institute and the Director of the East China Branch of the National College Institute of Manufacturing Automation.

# Contents

# Preface

The importance of solid waste management in addressing resource scarcity and environmental protection, and even in raising environmental awareness, is undeniable. The rapid emergence and application of new technologies, processes, and materials have brought about profound changes in industrial production and human life. Accordingly, the connotations, methods, and means of solid waste management have also undergone profound changes. People have also developed many new understandings of solid waste management.

Aimed at the whole process of solid waste management, this book is a collection of articles on the recent advances, new trends, and applications of solid waste management by academicians, researchers, and practicing engineers in the past several years. The book includes eight chapters and encompasses multidisciplinary areas within solid waste treatment ranging from reviews and methods to case studies. *Solid Waste Management – Recent Advances, New Trends and Applications* is a useful resource for professionals and managers in solid waste management, including undergraduate and postgraduate students, academicians, researchers, and practicing engineers.

**Pengzhong Li**
School of Mechanical Engineering,
Tongji University,
Shanghai, China

Chapter 1

# A Prefatorial View of Solid Waste Management

*Boniface Yeboah Antwi and Ammishaddai Ofori-Nyarko*

## Abstract

Solid waste management is the process of collecting, transporting, processing, and disposing of waste in a responsible and sustainable manner. Proper waste management is essential for public health, environmental protection, and the efficient use of resources. In many parts of the world, solid waste management remains a major challenge due to inadequate infrastructure, lack of funding, and limited awareness. The accumulation of solid waste in urban areas can lead to various health and environmental problems, including air and water pollution, disease transmission, and greenhouse gas emissions. Governments, organizations, and communities are implementing various waste management strategies to address these challenges, including waste reduction, recycling, composting, and waste-to-energy technologies. These strategies aim to reduce the amount of waste generated, recover valuable resources from waste, and minimize the negative impacts of waste on public health and the environment. Effective solid waste management requires a comprehensive and united approach that involves all stakeholders, including government, industry, civil society, and the public. Through collective action, we can create a more sustainable future by reducing waste, conserving resources, and protecting the environment.

**Keywords:** solid wastes, segregate, reduce, reuse, recycle, upcycle

## 1. Introduction

Waste management has a long and varied history, dating back to ancient times. The earliest forms of waste management were simple methods like collecting garbage and burying it in open pits on the outskirt of towns and villages and dumping it in water bodies. In ancient Rome, waste was collected from households by private contractors, and then dumped in designated sites outside the city walls. Another method was burning garbage in open-air pits, a practice that continued until the early twentieth century [1]. In the Middle Ages, waste products were often dumped into rivers or lakes, leading to severe pollution. During the Industrial Revolution, cities experienced a surge in population, which led to the growth of waste production. As a result, public health concerns emerged, and governments began to regulate waste disposal practices [2].

The modern era of waste management began in the late 1800s, with the creation of the first municipal waste disposal systems. These systems involved collecting and transporting waste to a central location, where it was sorted and recycled [3]. In the

1960s, the environmental movement led to increased awareness of the impact of waste on the environment, and recycling programs were introduced in some cities [4]. In the 1970s, the US Environmental Protection Agency (EPA) was established, and the Resource Conservation and Recovery Act (RCRA) was passed, which set standards for hazardous waste management [5]. In the 1980s and 1990s, waste-to-energy technologies were developed, which converted waste into energy through incineration or gasification [6].

Any material that is discarded after primary use is considered waste material. The increasing quantity and complexity of modern waste pose a serious risk to human health and the environment [7]. The increase in municipal solid waste in the modern world is a result of the world gearing toward urbanization. Waste generation is a burden that comes with development. Currently, municipal waste generated globally is estimated to be about two billion tons, and only 33% of that is managed properly [8]. The remaining 67% ends up in landfills, dumpsites, and the oceans then contributes to polluting our environment [7]. Reported that, by 2025, there will be about 4.5 billion urban residents generating 1.4 kg/capita/day of municipal waste.

Municipal solid wastes come from various sources, from households, offices, institutions, and other commercial organizations. The composition of solid waste from an area varies according to the lifestyle and living standards of the inhabitants of that area [3]. Lower to middle-income class generates more organic waste while high-income residents produce more metal, paper, and glass waste [9]. Solid wastes are made up of organic and inorganic materials, biodegradable and non-biodegradable materials. Because of the varying composition, strict and careful management practices are needed to control the flow of solid wastes from their source. Poorly managed waste pollutes the environment and oceans. It clogs drainage systems to cause flooding, increases the transmission of diseases, and air pollution from the burning of waste.

Safe handling of waste till final disposal is essential for any modern community or municipality. Proper handling of municipal waste could be an economic resource with great potential in the energy and manufacturing fields [9]. Effective waste management is a hands-on situation and requires participation from everyone, from the authorities to the individual waste generators [10]. In most developed countries, wastes are carefully regulated through well-established record-keeping systems. These countries have legislations in place to prevent improper disposal of waste that may cause pollution and outbreak of diseases. In developing countries, insufficient resources, and a lack of clear definitions, roles, responsibilities, and quality data have made waste management difficult [3]. Some of these developing countries have legislations to regulate solid waste management but have struggled to enforce these regulations.

As the realization that most of our resources are limited, it is necessary that we derive economic value from waste materials [11]. The process of finding value out of waste is known as recycling [12]. There are various ways in which waste materials can be recycled; the raw materials could be processed and used for another product or the waste material can be repurposed to generate energy. In order to extract value from waste, we must first start by properly managing solid waste right from the source.

Solid waste management includes; the generation, collection, treatment, disposal, and monitoring of waste. There are several factors such as environmental, political, economic, technical, and legislation that are taken into consideration during waste management. Sustainable waste management looks to promote reducing the amount of waste generated and recycling the majority of the waste. Minimizing waste produced heavily relies on the actions of the waste generator [12]. It is important that education and awareness in the area of waste management be prioritized. We shed light on the process of solid waste management, its importance, and its challenges.

## 2. Waste collection

The collection of waste is the first step to effective solid waste management. This includes collecting waste from the source and transporting it to be treated and/or disposed of. The waste collection process starts with generating, source segregation, storage, and transportation [13]. Collection and transporting of waste is the most critical and cost-effective part of solid waste management because it is laborious and requires the use of vehicles [11]. Factors ranging from the route to transport waste and the availability of storage for waste are all to be considered. It is a multipart process and requires intricate logistics management in order to operate well for a long time [14].

The collection system for waste depends on the source and type of waste. For instance, the collection system of waste from an urban area would not work in a rural area because of the difference in the amount of waste generated per a given time. Another scenario to look at is that waste from residential areas would be different from waste from commercial or industrial areas hence, cannot be collected the same way. Some areas generate more perishable waste products than others, hence the need for them to be collected and transported frequently. The type and quantity of waste produced will determine the scale and nature of the collection process. The collection of recyclable wastes has become very essential to most societies looking toward sustainable waste management.

In most urban areas the collection of waste is the responsibility of local authorities and in some areas, there are no formal methods of waste collection. Optimal locations for waste collection are a fundamental aspect of the cost-effectiveness and convenience of solid waste management [15].

## 3. Waste segregation

One modern system of waste collection is segregation at the source in order to minimize the complication of sorting through several heaps of waste before treatment. Waste segregation is considered to be an effective way for initiating the recycling process [16]. Source segregation is defined as separating 'useful' waste materials from the waste stream right after the waste has been generated. Waste is segregated either by recyclability or degradability. Ideally, waste should be segregated based on whether the material is paper, glass, organic, or plastic.

In doing this, the amount of waste that goes to landfills is reduced [16]. Source segregation of waste increases the amount of recycling, reuse, and recovery. When recyclable waste is not segregated at the source and is mixed with organic matter it makes sorting and recycling very difficult. In addition, the moisture content in waste is less when it is segregated at the source, this makes it easy for disposal by incinerating. Burning of recyclable wastes like plastics produces harmful gasses that harm the environment and its inhabitants [17].

Waste segregation begins at the basic source of waste like households, schools, offices, etc. The common method of segregating waste is to get different bins for the different waste types. These bins are labeled and placed conveniently so that after primary use, waste materials are stored in their appropriate bin. When the time comes for the waste to be transported, recyclable wastes are taken to the recycling sites while non-recyclable waste is taken to be disposed of. This helps to achieve a better recirculating rate for waste management systems [18]. The quality of raw materials for recycling is much higher in source-segregated waste than it is in materials sorted from a mixed waste stream.

It is important that all solid wastes generated must be stored properly while waiting to be transported. Plastic dustbins are the most commonly used waste storage devices in this part of the world. The bin must have a tight lid to prevent fumes from polluting the environment and also diseases from breeding from it. These bins should also be conveniently placed and accessible to waste generators and collectors.

## 4. Waste transportation

Transporting waste is the next step in solid waste management after storage. Waste materials are either transported to a treatment facility or a landfill to be disposed of. Transporting waste is a very complex and cost-inducive process. It requires a collection crew, vehicles, pick-up points, and intricate logistics [19]. In most urban and residential homes, the storage of waste is done by individuals and residents whereas the transportation is outsourced to private collectors by municipal authorities. In some instances, the waste is collected from bins placed in front of the respective residence for collection. Other situations require that the waste be sent to designated pick-up spots for collection. After waste is collected, they are transported to treatment sites to be sorted and treated before disposal. The main thing that must be looked at when analyzing waste transportation is economics [20]. It ranges from the cost of fuel for transporting to the cost of crew members collecting waste materials. The distance of transportation routes and the size of the axle of waste trucks contribute to the total cost of collecting and transporting waste [13]. If the distance between the collection sites and the treatment site is too far, the vehicles use up more fuel to make the trip. The capacity of the axle of the waste truck also contributes to the cost of transportation [20]. If the waste truck has a smaller axle then it must make multiple trips within a given week or month. A model adopted by some developed countries is that waste is moved by smaller vehicles to a temporary dumpsite, then large garbage trucks collect the solid waste from the temporary site to their preferred destination [21].

## 5. Waste treatment

Treatment of waste is the next essential part of solid waste management. In an ideal system where waste is segregated at the source, after collection, recyclable materials are sent to the appropriate recycling plants, and the remaining waste materials are sent to a landfill for treatment and disposal. However, in most cases, the segregation of waste is done after it has been transported to landfills. After recyclable materials are sorted from the waste stream the remaining waste materials have to be treated to either facilitate easy disposal or to be used as a resource material. Waste treatment varies according to the type of waste material and it includes, biotreatment, incineration, energy recovery, and pyrolysis, among others [22].

### 5.1 Biotreatment

Biotreatment is the preferred method for treating biodegradable waste materials. Biotreatment of solid waste materials is mainly by anaerobic degradation; this is a process where bacteria are used to break down components of organic matter. The process of anaerobic degradation is in four stages namely: hydrolysis, acidogenesis, acetogenesis, and methanogenesis [23]. The goal of this process is to generate renewable energy in the form of methane gas from organic solid waste [24]. Aerobic degradation is the preferred method for the treatment of large volumes of wastewater and sludge with rich organic matter because the process requires oxygen [22]. The difference between this and the former is that, anaerobic digestion happens in the absence of oxygen. Anaerobic degradation is the most extensively used form of biodegradation because of its versatility. It can be used to digest a large variety of biomaterials (sewage sludge, municipal solid wastes, industrial wastes, manure, etc.); the system can be used to produce energy generation and soil fertilization [25].

### 5.2 Incineration

The thermal treatment process is mostly reserved for combustible and dry waste materials. It involves burning the waste materials in a controlled environment. Most incineration facilities are known as waste-energy facilities; the generated heat and electricity can be supplied to small towns and other urban areas [26]. Incineration is the best treatment method for eliminating infectious components of waste [22]. It can reduce about 90% of the volume of municipal solid wastes and minimize dependency on fossil fuels [27]. Relative to other waste treatment methods such as composting and landfilling, incineration with air emission management systems is seen to be the best option [28].

Traditional incineration is known to emit greenhouse gasses and poses a health risk to the nearest inhabitants [29]. Traditional incineration is where by there are no air emission management systems hence, the smoke from the incinerator goes into the atmosphere. Incineration plants require high operation costs and are more expensive than controlled landfilling [30]. Because of this, waste generators may find alternatives to disposal than incinerating.

Other thermal treatment processes for solid wastes are gasification and pyrolysis. Biomass gasification is a thermochemical process whereby waste is heated at high temperatures with low oxygen to break it down for the production of syngas and char [31]. Syngas can be used as fuel and raw materials for the production of solid and liquid fuels [32]. Pyrolysis is a thermal degradation process whereby waste is heated at high temperatures in the absence of air to produce valuable materials such as char, wax, and fuel gasses [33]. Both gasification and pyrolysis are waste-to-energy methods of processing solid wastes and are sustainable way to foster a circular economy. However, both processes require a high amount of energy for processing and so have high operation costs [34].

### 5.3 Recycling

Recycling of waste material has become a necessity in modern-day waste management. With the continued increase in the volume of municipal solid wastes, it is important that most waste materials are redirected from landfilling. Recycling is any means by which waste material is converted into useful products either for their original or other purposes [35]. The recycling process begins with the waste generator; as stated already, sorting solid waste from the source is the best way to initiate the recycling process. The municipal solid waste stream is not homogenous hence recycling is difficult. Some recyclable waste materials are plastics, paper, glass, and metal. Recycling is the most practiced reduction technique for solid waste management. Some ways of recycling solid waste materials are described in the subsections.

#### *5.3.1 Recycling of plastics*

Plastic recycling is a globally recognized process of managing plastic waste. The versatility of polymer has made plastics integral in today's society, from packaging materials to medical and electronic equipment [36]. The most common kind of plastic waste is plastic packaging because it is mostly single-use [37]. Ref. [36] reported that plastics are recycled chemically or mechanically. Mechanical recycling is where plastic wastes are cleaned, melted, and extruded to form various products. These products usually have low quality and molecular weight than the primary products [37]. For chemical recycling, the plastics wastes go through several steps to break the long chain bonds in order to form new plastics or use them for fuel/other chemical products [38]. Not all plastics can be recycled both mechanically and chemically the recycling techniques are dependent on the type of plastic [36].

#### *5.3.2 Recycling of metals*

Recycling metals from solid waste streams is another huge step toward sustainable waste management and a circular economy. Light metals used for machine parts and food packaging are the most common in municipal waste streams [39]. Metal parts, aluminum, and steel cans and sheets pose a danger to the environment when they are dumped in landfills. While modern society looks at green environment and waste reduction, waste metals are better managed by recycling. There are a variety of methods in metal recycling including mechanical, hydrometallurgical, and pyrometallurgical processes [40]. Mechanical recycling of metals includes sorting the metals from the waste stream, processing them through mechanical processes such as shredding, and then melting the metal down in order to produce new products. This is a common recycling method for aluminum cans, steel sheets, and other scrap metals [41]. For pyrometallurgical recycling, the metal is heated at elevated temperatures to remove any impurities and other materials, then the resulting molten metal is cast into a new product. This process is mostly used to recycle lead, copper, and zinc [42]. The other widely used process for metal recycling is hydrometallurgy. This involves dissolving the metal in a chemical solution to recover/extract metals then the metal is processed into a new product. This technique is mostly used for precious metals like gold, silver, and platinum [43]. Other methods for recycling metals like electrochemical and bio metallurgical methods are being developed [44, 45]. The choice of recycling method is reliant on the kind type and quantity of waste metal as well as the environmental and economic impact [43].

#### *5.3.3 Recycling of glass*

Glass recycling involves melting down glass into molten form then it is refined to remove impurities before a new product is formed with the molten glass [46]. This process usually requires about 1500°C which is less energy than what is required to make new glass from raw materials [47]. There are two main ways by which glass is recycled. They are either by the closed-loop system where the recycled glass is used to make new products of the same type and color. With this method, there is very little loss in the quality of the glass [48]. The other method by which glass is recycled is the open-loop system whereby glass is recycled into new products that are not necessarily the same type or color as the original material [49]. An example is when glass is recycled to make fiberglass insulation or decorative tiles. The closed-loop system of recycling is considered the most sustainable because it conserves energy and releases minimal greenhouse gasses; open-loop recycling allows for a variety of materials to be made from recycled glass [48].

#### *5.3.4 Recycling of paper*

Paper is also one of the most common waste materials. It is used everywhere from offices to schools, for packaging to decorating [50]. After the waste paper is collected and taken to a recycling facility, the papers are sorted according to their types i.e., office paper, cardboard, etc. Then, the papers are shredded into smaller pieces and mixed with water to form a pulp. The pulp is screened thoroughly and refined to remove any impurities and/or non-fibrous materials. Chemicals like surfactants, caustic soda, and peroxides are then used to de-ink the paper [51]. The pulp is now bleached to get the desired brightness before it is repulped to make new materials like cardboard, tissue paper, or printing paper [52]. Paper recycling preserves natural resources and reduces the amount of waste sent to landfills.

Upcycling is another sustainable way to manage waste. It is when waste products are given a second life as it is upgraded and reused for another purpose [53]. Upcycling is about creating a new design and meaning to a material that has already served its purpose. The creation of art from plastic waste and making flower beds out of car tires and other waste containers can be considered upcycling. Waste materials are also recycled in the construction industry as building aggregates/materials [54]. Researchers have studied the effects of waste glass and plastics as aggregates and binders in concrete and cement materials. Studies show that if managed carefully, solid wastes can be effectively recycled into concrete and other cement materials without destroying the integrity of said materials [55].

### 5.4 Landfilling

Landfills are sites designed to store garbage; they are mostly used to store non-hazardous waste [56]. Waste in landfills undergo biological, physical, and chemical transformations [9]. Hence, extra steps must be put in place to prevent pollution and contamination of the environment. Sustainable landfilling is a concept to minimize the environmental impact of landfills while allowing for solid waste disposal [57]. Sustainable landfilling can be achieved by the use of landfill liners, energy recovery, gas and leachate collection, and frequent maintenance [9]. Gas emissions from municipal landfills can be collected to generate waste. Methane gas from composting waste can be captured and used to generate fuel or electricity either by steam turbines or oxide fuel cells [58]. Organic food waste can be secluded from other waste in landfills for composting to produce nutrient-rich soil [59]. By implementing the necessary practices, sustainable landfilling can minimize the environmental impact of solid waste disposal and offer economic breakthroughs.

## 6. Advantages of solid waste management

Solid waste management is an essential part of the development of any town, city, or urban area. Some of the key benefits of an effective solid waste management practice are: Improved public health, environmental protection, resource conservation, economic benefits, reduced greenhouse gas emissions, and improved esthetics.

Proper waste management practices help prevent the spread of diseases, reduce the risk of contamination, and minimize exposure to harmful toxins and pollutants. Poor solid waste management is linked with poor public health and affects the sustainable development of most cities [60].

Proper disposal of waste reduces the risk of soil, air, and water pollution, preserving the environment and protecting wildlife [9].

Solid waste management involves the recovery and recycling of valuable resources, such as metals, plastics, and paper. This reduces the amount of waste sent to landfills and deductively conserve natural resources [61].

Proper waste management practices can create jobs and generate income by collecting, transporting, and processing waste materials [62].

Landfills are a significant source of greenhouse gas emissions, particularly methane. Proper solid waste management, including waste reduction, recycling, and composting, can help lower these emissions and mitigate the effects of climate change [48].

Properly managed solid waste helps maintain a clean and tidy appearance in public spaces, reducing litter and improving the overall quality of life in a community.

## 7. Challenges with municipal solid waste management

Some challenges faced by most municipalities when enforcing solid waste management are: Increased waste generation, inadequate infrastructure, financial constraints, public attitudes and behavior, recycling and composting, and hazardous waste management.

The sheer volume of waste generated by urban areas is steadily increasing each year, and finding suitable locations for landfills and waste disposal facilities can be difficult due to limited space and environmental regulations [9].

On the infrastructure, many cities lack the necessary systems and resources to effectively manage waste, including collection and transportation channels, sorting and processing facilities, and trained personnel [13].

Economically, implementing effective waste management systems can be costly, and many municipalities may lack the necessary funding to invest in these systems [63].

The public behavioral limitations originates from the poor waste management practices learnt from less informed individuals who illegally dump or litter the environment and create long term challenges for municipalities with consequential undermine of their efforts at managing the waste [64].

Recycling and composting can help reduce the volume of waste sent to landfills, however, these activities can be challenging to implement effectively due to logistical and economic barriers, such as inadequate recycling infrastructure and limited markets for recycled materials [63].

Managing hazardous wastes, such as chemicals, batteries, and electronic waste, present unique challenges and requires specialized facilities and processes to ensure proper disposal and harmless to the environment and public health [65].

## 8. Conclusion

For the sustainable development of our cities and urban areas, solid waste management must be practiced effectively. Best solid waste management programs improve the health and environmental conditions of residents and animals. This creates an opportunity for waste materials to be used as a resource which in turn, generates significant economic benefits. Recycling provides job opportunities for several people and enhances the economic quota of a particular region and/or country.

The extraction of energy from waste material is an important addition to the solutions for the global depletion of non-renewable resources. Proper management of waste reduces the volume of waste on landfills and subsequently minimizes the emissions of greenhouse gasses. A significant reduction in waste generation from successful solid waste management practices is a huge step toward achieving sustainable development goals.

Despite the many challenges that are faced with the implementation of solid waste management, it is necessary that municipalities and governments provide the right infrastructure and systems that will minimize waste, channel generated waste to recycling, and minimize landfilling to protect the people and the environment.

## Author details

Boniface Yeboah Antwi* and Ammishaddai Ofori-Nyarko
Council for Scientific and Industrial Research-Institute of Industrial Research, Accra, Ghana

*Address all correspondence to: byantwi@csir.org.gh

## References

[1] Barles S. History of waste management and the social and cultural representations of waste. In: The Basic Environmental History. Cham, Switzerland: Springer International Publishing; 16 Oct 2014. pp. 199-226

[2] Buck HJ. Should carbon removal be treated as waste management? Lessons from the cultural history of waste. Interface Focus. 2020;**10**(5):20200010

[3] Chandrappa R, Das DB. Solid waste management: Principles and practice. Heidleberg: Springer Science & Business Media; 2012. pp. 363-369

[4] Wilson DC, Rodic L, Modak P, Soos R, Carpintero A, Velis K, et al. Global Waste Management Outlook. Leeds: Sympletic Publications, UNEP; 2015

[5] Lee C. Toxic waste and race in the United States. In: Race and the Incidence of Environmental Hazards. New York: Routledge; 2019. pp. 10-27

[6] Mukherjee C, Denney J, Mbonimpa EG, Slagley J, Bhowmik R. A review on municipal solid waste-to-energy trends in the USA. Renewable and Sustainable Energy Reviews. 2020;**119**:109512

[7] Daniel H, Perinaz BT. What a waste: A global review of solid waste management. Urban Development Series. Knowledge Papers no. 15. © World Bank, Washington, DC. 2012. Available from: http://hdl.handle.net/10986/17388. License: CC BY 3.0 IGO

[8] Atlas W. Waste Atlas-Interactive Map With Visualized Waste Management Data. (nd). Available from: (d-waste.com)

[9] Nanda S, Berruti F. Municipal solid waste management and landfilling technologies: A review. Environmental Chemistry Letters. 2021;**19**:1433-1456

[10] Asnani PU. India Infrastructure Report: Ch. 8 Solid Waste Management. New Delhi, India: Oxford University Press; 2008. pp. 160-189

[11] Beliën J, De Boeck L, Van Ackere J. Municipal solid waste collection and management problems: A literature review. Transportation Science. 2014;**48**(1):78-102

[12] Demirbas A. Waste management, waste resource facilities and waste conversion processes. Energy Conversion and Management. 2011;**52**(2):1280-1287

[13] Das S, Bhattacharyya BK. Optimization of municipal solid waste collection and transportation routes. Waste Management. 2015;**43**:9-18

[14] Hannan MA, Lipu MH, Akhtar M, Begum RA, Al Mamun MA, Hussain A, et al. Solid waste collection optimization objectives, constraints, modeling approaches, and their challenges toward achieving sustainable development goals. Journal of Cleaner Production. 2020;**277**:123557

[15] Lu JW, Chang NB, Liao L, Liao MY. Smart and green urban solid waste collection systems: Advances, challenges, and perspectives. IEEE Systems Journal. 2015;**11**(4):2804-2817

[16] Otitoju TA, Seng L. Municipal solid waste management: Household waste segregation in Kuching South City, Sarawak, Malaysia. American Journal of Engineering Research (AJER). 2014;**3**(6):82-91

[17] Susanth GS, Livingston LJ, Livingston LA. Garbage waste segregation using deep learning techniques. In: IOP Conference Series: Materials Science and Engineering. Bristol: IOP Publishing; 2021;**1012**(1):012040

[18] Rousta K, Ordoñez I, Bolton K, Dahlén L. Support for designing waste sorting systems: A mini review. Waste Management & Research. 2017;**35**(11):1099-1111

[19] Singh GK, Gupta K, Chaudhary S. Solid waste management: Its sources, collection, transportation and recycling. International Journal of Environmental Science and Development. 2014;**5**(4):347

[20] Brogaard LK, Christensen TH. Quantifying capital goods for collection and transport of waste. Waste Management & Research. 2012;**30**(12):1243-1250

[21] Ghose MK, Dikshit AK, Sharma SK. A GIS based transportation model for solid waste disposal–A case study on Asansol municipality. Waste Management. 2006;**26**(11):1287-1293

[22] Hamer G. Solid waste treatment and disposal: Effects on public health and environmental safety. Biotechnology Advances. 2003;**22**(1-2):71-79

[23] Meegoda JN, Li B, Patel K, Wang LB. A review of the processes, parameters, and optimization of anaerobic digestion. International Journal of Environmental Research and Public Health. 2018;**15**(10):2224

[24] Bátori V, Åkesson D, Zamani A, Taherzadeh MJ, Horváth IS. Anaerobic degradation of bioplastics: A review. Waste Management. 2018;**80**:406-413

[25] Kang AJ, Yuan Q. Enhanced anaerobic digestion of organic waste. In: Solid Waste Management in Rural Areas. Rijeka: Intechopen; 2017. p. 7

[26] Makarichi L, Jutidamrongphan W, Techato KA. The evolution of waste-to-energy incineration: A review. Renewable and Sustainable Energy Reviews. 2018;**91**:812-821

[27] Alam O, Qiao X. An in-depth review on municipal solid waste management, treatment and disposal in Bangladesh. Sustainable Cities and Society. 2020;**52**:101775

[28] Li HY, Gao PP, Ni HG. Emission characteristics of parent and halogenated PAHs in simulated municipal solid waste incineration. Science of the Total Environment. 2019;**665**:11-17

[29] Cudjoe D, Acquah PM. Environmental impact analysis of municipal solid waste incineration in African countries. Chemosphere. 2021;**265**:129186

[30] Rand T, Haukohl J, Marxen U. Municipal Solid Waste Incineration. A Decision Maker's Guide. Washington, DC: The International Bank for Reconstruction and Development, World Bank; 2000. pp. 1-11

[31] Molino A, Chianese S, Musmarra D. Biomass gasification technology: The state of the art overview. Journal of Energy Chemistry. 2016;**25**(1):10-25

[32] Kobayashi M. Dry Syngas Purification Processes for Coal Gasification Systems. Journal of Japan Institute of Energy. Tokyo. 2021. pp. 1-49

[33] Chen D, Yin L, Wang H, He P. Pyrolysis technologies for municipal solid waste: A review. Waste Management. 2014;**34**(12):2466-2486

[34] Jaroenkhasemmeesuk C, Tippayawong N. Technical and economic analysis of a biomass pyrolysis plant. Energy Procedia. 2015;**79**:950-955

[35] Kassim SM. The importance of recycling in solid waste management. In: Macromolecular Symposia. Weinheim: WILEY-VCH Verlag; 2012;**320**(1):43-50

[36] Rahimi A, García JM. Chemical recycling of waste plastics for new materials production. Nature Reviews Chemistry. 2017;**1**(6):0046

[37] Ragaert K, Delva L, Van Geem K. Mechanical and chemical recycling of solid plastic waste. Waste Management. 2017;**69**:24-58

[38] Thiounn T, Smith RC. Advances and approaches for chemical recycling of plastic waste. Journal of Polymer Science. 2020;**58**(10):1347-1364

[39] Gesing A, Wolanski R. Recycling light metals from end-of-life vehicle. Journal of Metals. 2001;**53**(11):21-23

[40] Panayotova M, Panayotov V. Review of Methods for the Rare Earth Metals Recycling. Sofia: University of Mining and Geology; St. Ivan Rilski Publishing House; 2012;**55**:142-147

[41] Latif A, Ingarao G, Gucciardi M, Fratini L. A novel approach to enhance mechanical properties during recycling of aluminum alloy scrap through friction stir consolidation. The International Journal of Advanced Manufacturing Technology. 2022;**119**(3-4):1989-2005

[42] Bellemans I, De Wilde E, Moelans N, Verbeken K. Metal losses in pyrometallurgical operations-A review. Advances in Colloid and Interface Science. 2018;**255**:47-63

[43] Asadi Dalini E, Karimi G, Zandevakili S, Goodarzi M. A review on environmental, economic and hydrometallurgical processes of recycling spent lithium-ion batteries. Mineral Processing and Extractive Metallurgy Review. 2021;**42**(7):451-472

[44] Su X. Electrochemical separations for metal recycling. The Electrochemical Society Interface. 2020;**29**(3):55

[45] Ilyas S, Lee JC. Biometallurgical recovery of metals from waste electrical and electronic equipment: A review. ChemBioEng Reviews. 2014;**1**(4):148-169

[46] Dyer TD. Glass recycling. In: Handbook of Recycling. Amsterdam: Elsevier; 2014. pp. 191-209

[47] Lebullenger R, Mear FO. Glass recycling. In: Springer Handbook of Glass. Cham, Switzerland: Springer; 2019. pp. 1355-1377

[48] Ran W, Chen F, Wu Q, Liu S. A study of the closed-loop supply chain coordination on waste glass bottles recycling. Mathematical Problems in Engineering. 2016;**2016**:9. Article ID 1049514. DOI: 10.1155/2016/1049514

[49] Lessard JM, Habert G, Tagnit-Hamou A, Amor B. Tracking the environmental consequences of circular economy over space and time: The case of close-and open-loop recovery of postconsumer glass. Environmental Science & Technology. 2021;**55**(17):11521-11532

[50] Čabalová I, Kačík F, Geffert A, Kačíková D. The effects of paper recycling and its environmental impact. Environmental Management in Practice. 2011;**17**:329-350

[51] Mauchauffé R, Lee SJ, Han I, Kim SH, Moon SY. Improved de-inking of inkjet-printed paper using environmentally friendly atmospheric pressure low

temperature plasma for paper recycling. Scientific Reports. 2019;**9**(1):14046

[52] Ervasti I, Miranda R, Kauranen I. A global, comprehensive review of literature related to paper recycling: A pressing need for a uniform system of terms and definitions. Waste Management. 2016;**48**:64-71

[53] Wegener C. Upcycling. In: Glăveanu, VP, Tanggaard L, Wegener C, editors. Creativity — A New Vocabulary. Palgrave Studies in Creativity and Culture. London: Palgrave Macmillan; 2016. pp. 181-188

[54] Tang Z, Li W, Tam VW, Xue C. Advanced progress in recycling municipal and construction solid wastes for manufacturing sustainable construction materials. Resources, Conservation & Recycling: X. 2020;**6**:100036

[55] Okeyinka OM, Oloke DA, Khatib JM. A review on recycled use of solid wastes in building materials. International Journal of Civil and Environmental Engineering. 2015;**9**(12):1578-1587

[56] Ye J, Chen X, Chen C, Bate B. Emerging sustainable technologies for remediation of soils and groundwater in a municipal solid waste landfill site--A review. Chemosphere. 2019;**227**:681-702

[57] Parameswari K, Al Aamri AM, Gopalakrishnan K, Arunachalam S, Al Alawi AA, Sivasakthivel T. Sustainable landfill design for effective municipal solid waste management for resource and energy recovery. Materials Today: Proceedings. 2021;**47**:2441-2449

[58] Moya D, Aldás C, López G, Kaparaju P. Municipal solid waste as a valuable renewable energy resource: A worldwide opportunity of energy recovery by using waste-to-energy technologies. Energy Procedia. 2017;**134**:286-295

[59] Raksasat R, Lim JW, Kiatkittipong W, Kiatkittipong K, Ho YC, Lam MK, et al. A review of organic waste enrichment for inducing palatability of black soldier fly larvae: Wastes to valuable resources. Environmental Pollution. 2020;**267**:115488

[60] Abubakar IR, Maniruzzaman KM, Dano UL, AlShihri FS, AlShammari MS, Ahmed SM, et al. Environmental sustainability impacts of solid waste management practices in the global south. International Journal of Environmental Research and Public Health. 2022;**19**(19):12717

[61] Kumar S, Smith SR, Fowler G, Velis C, Kumar SJ, Arya S, et al. Challenges and opportunities associated with waste management in India. Royal Society Open Science. 2017;**4**(3):160764

[62] Pereira TD, Fernandino G. Evaluation of solid waste management sustainability of a coastal municipality from northeastern Brazil. Ocean & Coastal Management. 2019;**179**:104839

[63] Das S, Lee SH, Kumar P, Kim KH, Lee SS, Bhattacharya SS. Solid waste management: Scope and the challenge of sustainability. Journal of Cleaner Production. 2019;**228**:658-678

[64] Awasthi AK, Cheela VS, D'Adamo I, Iacovidou E, Islam MR, Johnson M, et al. Zero waste approach towards a sustainable waste management. Resources, Environment and Sustainability. 2021;**3**:100014

[65] Yong YS, Lim YA, Ilankoon IM. An analysis of electronic waste management strategies and recycling operations in Malaysia: Challenges and future prospects. Journal of Cleaner Production. 2019;**224**:151-166

## Chapter 2

# Solid Waste Management

*Nazish Huma Khan, Nida Naz, Mohammad Nafees, Nida Gul and Tooba Saeed*


## Abstract

In many countries of the world, especially underdeveloped countries, the practice of solid waste management is inefficient. Solid waste management has become a difficult environmental issue. Due to poor waste handling practices, many environmental and health issues arise. In this regard, many countries are trying to find ways to deal with the problem of solid waste. This chapter is an overview of solid waste management practices knocking the waste minimization techniques that play an important role in eliminating environmental problems. In developing countries, the practice of waste handling for infectious and non-infectious waste is of mixed type. Such mismanagement of solid waste paves the way for environmental pollution, leading to adverse effects on human health. Various factors such as poor policies, inefficient organizations, lack of financial support and poor governance, are the major constrains in safe waste management. Therefore, it is considered difficult to manage the recovery and safe disposal of solid waste. This study shows that there should be an appropriate organizational configuration for the separate treatment of different types of solid waste. For this, the authorities concerned must be strengthened financially and in skilled manpower for a good management of solid waste with a good recovery of resources.


**Keywords:** solid waste, sources, health problems, waste disposal, waste management

## 1. Introduction

Majority countries of the world are facing the issue of solid waste management. Initially it was thought that all the waste coming from household and animal wastes are solid waste. Later on it was proved that all the waste generated as a result of catastrophic events as well as construction and demolition sites are considered to be solid waste [1]. But solid waste are any materials that are being generated as a result of our daily activities, such as packages, bottles, leftovers, newspapers, equipment, gadgets, batteries, and colors [2]. Solid Waste Management (SWM) is becoming a global challenge throughout the world because of overpopulation and change in consumption patterns [3]. Improper disposal of solid waste causes variouse environmental problems like soil, water and air pollution as well as surface and ground water contamination. Decomposition of organic waste causes greenhouse gases into atmosphere [4]. While many health issues are also related to improper disposal of solid waste. The management of solid waste requires interdisciplinary links between politics, planning, geography, economics, public health, sociology, engineering, and materials science [2]. Integrated solid waste management is a process which is completely based on 3Rs (Reduce, Reuse, Recycle), with the goal of reducing the amount of waste that must be disposed of while maximizing material and energy recovery from waste, **Figure 1** [1].

Growing economies and populations in emerging nations like China, India, and Brazil have produced enormous amounts of solid waste that need to be managed [7, 8]. A number of these nations desire an advanced sustainable solid waste management system that supports better source material sorting and high recycling rates, but they do not have the necessary SWM capabilities to balance their sustainable development objectives [9]. Among the developed countries, the amount of waste generated in the United States was 88.1 million tons in 1960, later on it was increased up to 250 million tons in 2008 [10]. While in developing countries, the practice of solid waste management is inefficient. In Pakistan, there is no adequate waste collection and disposal system. All types of waste (industrial, municipal, hospital, etc.) are treated in the same way. The solid waste generation rate in Pakistan is calculated at 70,715 tonnes per day. In total, about 51–69% of solid waste is collected daily in Pakistan, while the rest of the waste is burnt or illegally dumped [11].

Urbanization and overpopulation is the root cause of increasing waste. By 2025 there will be 8 billion of world's population reach, out of which 5 billion will be living

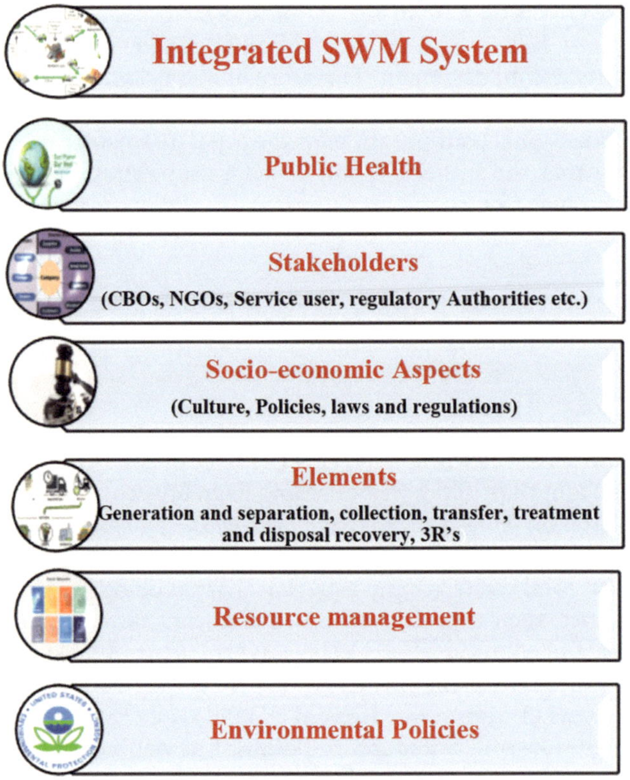


**Figure 1.**
*Integrated SWM system [5]. The poor policies of transportation, energy, and waste management causes one in eight deaths worldwide in 2012, being related to air pollution [6].*

in cities [12]. Similarly, the rapid population growth, ongoing urbanization, and the expansion of commercial, industrial, and service activities in Iran have increased waste production, which has caused a significant issues including environmental contamination [13]. Most MSWs in many nations are landfilled, while facilities for waste treatment including composting and incineration of wastes are rarely employed [14]. Solid waste management is important with the goals to plan strategies to address the health, environmental, esthetic, land-use, resource, and economic issues that arise from improper waste disposal [3]. The responsible team is municipal authority working at local level. Due to lacking of resources, the majority of municipalities do not collect the increasing level of waste. SWM is essential for conserving urban landscapes, people's health, and cities' reputations. It is important to support the waste colleting authority and all issues associated with the collection and disposal of solid waste because the situation will get worse over time due to the high rate of urbanization [15].

### 1.1 Types of solid waste

Solid waste is of different type, depending upon the sources from which it is generated, **Table 1** [4, 5]. These wastes carry two main types including hazardous and non-hazardous waste. Hazardous waste includes metals (Cd, Cr, Cu, Ni, Pb, etc.), chemicals, sharp objects, paints and is difficult to reuse or recycle. While non-hazardous waste includes food waste, paper, plastic, tin metals, glass, wood, etc. that can be recovered in a conventional solid waste management system [11]. The majority of waste generated by industries, hospitals and laboratories is considered hazardous while waste from commercial areas and municipal waste is not hazardous to health. Depending on the nature of the type of waste, treatments or appropriate disposal are required as a management strategy [5].

### 1.2 Adverse health impacts

The effects of solid waste on health may vary based on a number of factors, including the type of waste management practices, the traits and behaviors of the population exposed, the period of Interventions for exposure, prevention, and mitigation. The health consequences included mortality, neonatal outcomes, cancer, respiratory illnesses, gastroenteritis, vector-borne infections, cardiovascular disorders, and mental health issues [16]. Vegetable peels, common household waste,

| Sources | Industrial Area | Commercial Area | Residential Area | Institutional Area | Hospital Services |
|---|---|---|---|---|---|
| Types of Wastes | Hazardous and nonhazardous wastes, e-waste, metallic and non-metallic waste, paper, plastics etc. | Glass, metal, plastic, e-waste, hazardous waste etc. | Kitchen waste, textile, construction and demolition, batteries and hazardous waste etc. | Ashes, infectious and toxic waste, hazardous and e-waste, plastic and paper waste, glass waste etc. | Hazardous (sharps, medical waste), non-hazardous waste (food waste, paper, metal, plastic) Construction and demolition waste etc. |

**Table 1.**
*Sources and types of solid waste.*

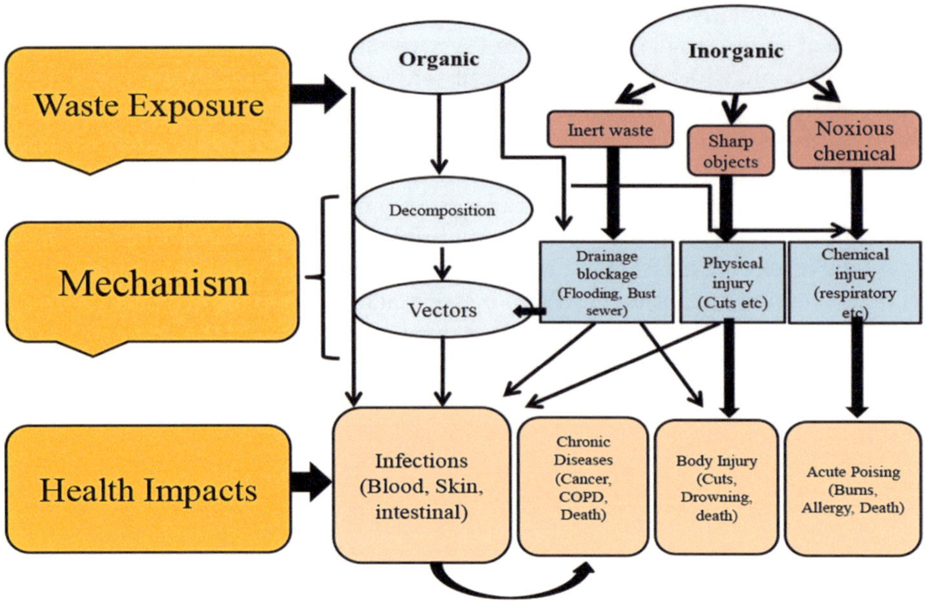


**Figure 2.**
*Poor SWM and adverse health impacts [17].*

medical waste, e-waste, etc. are all thrown away. Decomposition of the waste results in the release of numerous noxious gases, which draw numerous mice and other pests to the area and create major health dangers. Due to the unscientific and unsanitary way of life of the people of Garia, Kolkata, individuals frequently suffer from diseases like diarrhea, malaria, and dengue. While certain health effects may be rapid, clear to see, and directly related to the exposure to solid waste, others may be occult, long-lasting, and difficult to identify to a specific type of trash. The four groups-model represents the health implications easier to understand, **Figure 2** [17]. The model shows the mechanism of improper waste disposal and its health impacts on human health. Such diseases are infections and chronic effects. The responsible sources for health diseases are improper disposal of hazardous type of waste such as sharps etc.

## 2. Waste management practices

The adoption of integrated sustainable waste management techniques that are based on a comprehensive, multifaceted, and multi-stakeholder approach to the design and operation of MSW management systems. The 3Rs principles, prioritizing actions in accordance with the waste pyramid hierarchy, and embracing life-cycle thinking are all crucial components of this, **Figure 3**. This practice of waste management is quite easy and cheap. While being encouraged by contemporary waste management strategies, reuse, and recycling, composting, and safe disposal through landfills are usually not carried out. A significant fraction of waste generated in underdeveloped nations is not recycled. The only problem is the improper and unsafe disposal of unwanted items. It is challenging to recycle or compost due to the lack of waste recycling. Due to this, a large amount of solid waste in developing nations is dumped in open areas and frequently burned [17].

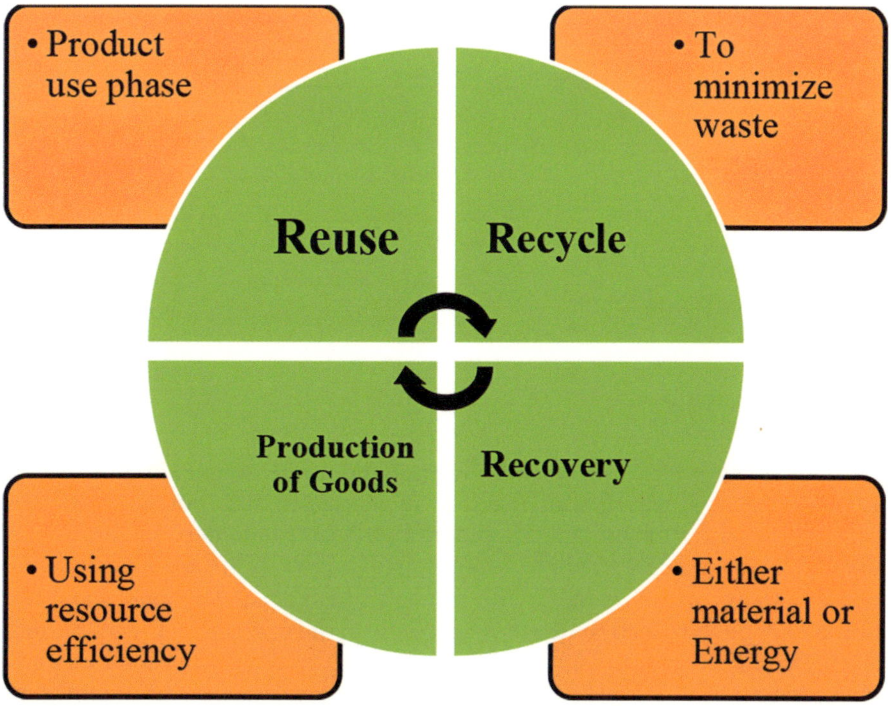


**Figure 3.**
*Safe disposal to minimized residual waste [18].*

The existence of laws and policies controlling waste management and the extent to which they are enforced, the amount of funds sources, and the type and amount of waste generated all have an impact on the differences in waste management methods. The final disposal is frequently at an open dumping site on the outskirts of the city, with collection frequently occurring near the source or temporary dumping site. Truckloads of wastes are usually dumped at dump sites, which are typically large open spaces or water channel. Dumped waste is searched for usable and recyclables items by the scavengers or it is frequently burned to minimize the bulk [17]. Solid waste composition is complex due to inadequate solid waste sorting at any level and may contain industrial, medical, electronic, and human waste deposited on the same open grounds where all the other municipal waste is dumped [19].

### 2.1 Landfill technology

The process of properly disposing of biodegradable and non-biodegradable wastes in a landfill (set apart and away from locality) is known as landfilling. In many nations, landfilling has been the traditional and most profitable method of waste disposal [20]. Due to its low cost and labor-intensiveness, landfilling is preferable to incineration and recycling of municipal solid waste. Additionally, by using its landfill gas and leachate for energy production, a consolidated landfilling can also generate income [20]. However, due to improvements in recycling, composting, incineration, and energy recovery technologies, landfilling of municipal solid waste decreased from 89% in 1980 to less than 53% in 2014 in the USA [20]. According to the type of waste, such as household waste, toxic chemicals, biohazards, biomedical

| Category | Type of Solid Waste Disposal |
| --- | --- |
| Class 01 | Soil disposal. |
| Class 02 | Disposal of construction and demolition, renovation waste and minerals. |
| Class 03 | Disposal of municipal solid waste. |
| Class 04 | Disposal of wastes coming from industries and commercial sectors. |
| Class 05 | Disposal of hazardous waste. |
| Class 06 | Underground disposal of wastes. |

**Table 2.**
*Categories of solid waste disposal.*

wastes, radioactive wastes, as well as building, demolition, and restoration wastes, different landfills and remediation facilities are located in different areas. There is no proper landfill system for the safe disposal of solid waste in underdeveloped countries such as Pakistan. The practice of solid waste disposal is in open land areas or in the nearby areas of water channels. This not only opens a pathway for soil and water contamination but also lead to severe health issue to the scavengers (birds, animals, human). There are the following categories according to the type of solid waste which is to be refused, **Table 2**. These types of waste area generated from different sources as mentioned in the table.

## 2.2 Bioreactor landfills

A bioreactor landfill is a specially designed contemporary landfill that transforms waste disposal from storage to treatment. In comparison to conventional landfills, bioreactor landfills have different advantages, i.e. (i) improved leachate quality, (ii) storage and partial in situ treatment of leachate, (iii) high landfill gas production rates and yields, (iv) effective gas recovery for on-site flaring, (v) early waste stabilization, (vi) improved biodegradation of biodegradable components in municipal solid waste leading to faster settlement, and (vii) cost- and time-effective (ix) increased waste into energy conversion. Temperature inside bioreactor ranges from 45 to 60°C. While the preferable temperature for the optimal activity of methanogenic bacteria range between 35 and 45°C. At a bioreactor landfill with pH levels between 6 and 8, and alkalinity levels about 2000 mg/L, the methanogenic bacterial population also thrives in alkaline conditions [20]. Bioreactor landfills can be divided into anaerobic, aerobic and semi-aerobic landfills.

1. **Anaerobic bioreactor landfills,** an anaerobic bacteria transform biodegradable wastes into volatile fatty acids, which are subsequently converted into landfill gases like $CH_4$ and $CO_2$. Organic wastes are degraded anaerobically to produce leachate and landfill gases by a number of sequential processes, including hydrolysis, acidogenesis, acetogenesis, methanogenesis, anaerobic oxidation, and fermentation [20].

2. **Aerobic bioreactor landfills** speed up the breakdown of waste by supplying aerobic bacteria with excess oxygen. The aerobic bacteria generate energy by oxidizing

organic molecules, which primarily produces $CO_2$ and water. Aerobic degradation happens more quickly than anaerobic degradation because aerobic bacteria reproduce more quickly as a result of aerobic respiration, which is more efficient at generating energy than anaerobic respiration. By adding air to the soil layers and waste matter, aerobic bioreactor landfills promote aeration. Methanogenesis and $CH_4$ synthesis in landfills are lowered from starting levels of 60 to 10% within 7–10 days while anaerobic bacteria development is slowed [20].

3. **Landfills with semi-aerobic bioreactors** provide partially oxygen-deficient environments that promote the growth of both aerobic and anaerobic bacteria. According to this, methanation, hydrolysis, and fermentation likely occur simultaneously, albeit depending on the oxygen level, aerobic and anaerobic reactions may compete. The corresponding aerobic and anaerobic microbes may also locate in the landfill's most advantageous microbiome niches. In semi-aerobic bioreactor landfills, air is transmitted through the waste layer from the bottom of the landfill, whereas air is injected into the aerial space of aerobic bioreactor landfills. Air can naturally circulate through the leachate collecting pipes in some designs [20].

## 3. Thermal chemical conversion

This is a process of conversion solid waste into energy by thermal decomposition of waste material [18]. There are various ways for treating solid waste, and choosing the best one relies on the kind of waste, the amount of land that is available, and the cost of disposal, **Figure 4** [4]. These are as follows.

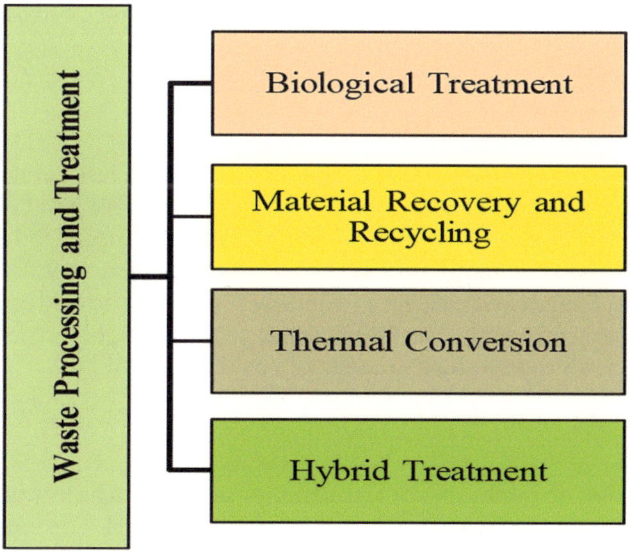


**Figure 4.**
*Treatment techniques of solid waste.*

### 3.1 Incineration

It is a controlled combustion process for producing gases and residue including non-combustible material by burning solid wastes at high temperatures of around 1000°C and above in the presence of excess air (oxygen) [4, 18]. The ability to employ the incine ration process to reduce the original amount of combustible MSW by 80–90% is one of its most alluring aspects [4]. Solid waste incineration typically produces 65 to 80 percent thermal energy. Stoker and fluidized bed furnaces are the two types that are most frequently used. Advantages: they are energy efficient, odorless as well as noiseless; disadvantages: Dioxin formation, SOx and NOx emissions, and particle creation; expensive; requires skills during its handling [18]. In undeveloped countries, majority of incineration plants are non-functional and hazardous waste is discarded and handled like non-hazardous type of waste.

### 3.2 Compaction

All the collected wastes are compressed and break into small pieces for further operations [4].

### 3.3 Pyrolysis

This is an endothermic process. Solid wastes are burned at the temperature range between 400 and 1000°C. Syngas, char and oil are produced as a result of pyrolysis [4] and a mixture of combustible gases CO, $CH_4$ and $H_2$ [18]. The feedstock for high-quality pyrolysis products should consist of a certain type of waste (plastic, rubber, electronics, electrical waste, wood waste, etc.). Several earlier research that focused more on the pyrolysis process itself than on the potential economic applications of the pyrolysis products reported on the pyrolysis of specific types of waste. Recently, pyrolysis has drawn interest in especially for recycling discarded tyres in order to recover oil, wire, carbon black, and gas.

### 3.4 Gasification

The temperature ranges between 1000 and 1400°C and less amount of oxygen of required [21]. Pyrolysis and gasification is more preferable and favorable process of treating MSW than incineration. Due to its lower impacts on environment. These processes use less energy for the reduction of waste up to 95% than incineration. Advatages: Syngas and slag are produced, while sulfuric acid is obtained from sulfur containing waste [18]. Disadvantages: it produces highly toxic polyhalogenated organics as a product, high amount of coal is required [18].

### 3.5 Composting

This is a process of aerobic biological decomposition of organic waste under controlled conditions, such as temperature, humidity, and pH. The indigenous microorganisms (thermophile and mesophile) transform organic matter into compost, a stable product [21]. The resulting compost functions as a soil conditioner and has uses in landscaping, agriculture, and horticulture [22]. Due to the gradual involvement of microorganisms during the breakdown process, composting is a "batch" process. Composting is a safe and ecofriendly process [21].

### 3.6 Hydrothermal carbonization

Carbonization is a relatively new thermal conversion method that turns solid waste into carbonaceous residues by heating it to a temperature between 180 and 350 C in a water-based environment (i.e., hydrochar). This process is exothermic in nature, moisture-rich waste materials are used, and energy requirements to run the device are low. But it is a costly method [18].

## 4. Waste management system

There are six different systems that are considered as waste management system, **Figure 5**:

1. Industrial Symbiosis/Inhouse waste handling: In this techniques, waste of one industry is used as a resource of production in another industry. For examples, waste wood of match industry is utilized as a resource in Chip-board industry. While waste of paper mill is used in packaging industry for packing of eggs, fruits and vegetables etc. [23].

2. Industrial waste management system

3. Littering/unmanaged waste Handling

4. Return system: used cans or bottles are returned to the company or industries for its reuse after the recycling of their respective waste (**Figure 6**).

5. Hazardous waste management system

6. Public or private Municipal Waste Management System [25].

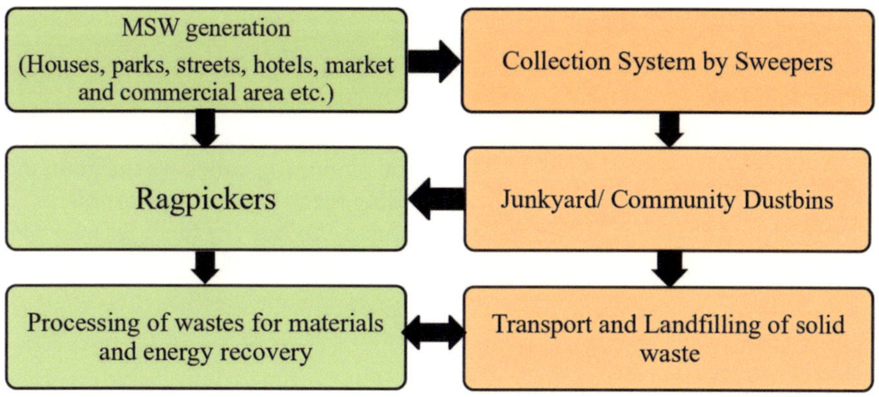


**Figure 5.**
*A model of WM system [5].*

**Figure 6.**
*SWM model for return system [24].*

## 5. Challenges to solid waste management (SWM)

There are the following challenges which come across the SWM.

### 5.1 Waste generation and characterization

Waste generation rate is directly depending upon the population of an area. People disposed of their waste openly or discarded inappropriately. The most meaningful way of waste management is to educate people to reduce waste. Because of the modern era industries and institutions are going towards modern food packaging which itself is a big problem leading towards the generation of waste [26]. Solid wastes are characterized in terms of corrosive, ignitability, reactive and toxicity [4] and are also categorized on the basis of biodegradable, nonbiodegradable, organic inorganic and natural wastes [26].

### 5.2 Lack of funding

Overpopulation has necessitated the provision of sophisticated infrastructure for urban areas, and landfill selection is essential. Dumping zones are the main issue in dumping sites and a lack of government funding for urban local governments (ULBs) to effectively manage generated solid waste. We lack the facilities necessary to conduct a suitable process for SWM, due to lack of funding [26].

### 5.3 Lack of awareness

Govt, do not taking entrust towards the awareness programs of locality and their participatory approach towards SWM. The modern techniques of SWM should be adopted on large scale by adopted nations [18].

### 5.4 Institutional setup

There is less interaction between the federal and state governments causes a delay in reporting information from one to the other, which delays proper implementation at the sub-national level. Major barriers include a lack of collaboration with urban municipal governments for specific action plans and a weak implementation of policy [26].

### 5.5 Segregation of waste

Solid wastes contain every type of material when they are openly dumped. We need properly managed and scientific system for their safe disposal. Improper waste segregation is hazardous and causes serious health problems [21].

### 5.6 Involvement of other sectors

Many organizational sectors were established to improve the process of collecting MSW and source segregation it, but due to societal ignorance these approaches are ignored [26]. Rag-pickers might be employed by the organized sector to increase MSW collection efficiency and source segregations. Nevertheless, this enormous potential has gone untapped because there aren't enough recycling industries and people do not accept recycling.

## 6. Solid waste management and sustainable development goals (SDGs)

The importance of maintaining health and protecting the environment through proper solid waste management in cities has increased because of seventeen Sustainable Development Goals (17SDGs). The SDG agenda promotes higher reuse and recycling as well as decreased waste generation. It mentions SDGs 3 (ensure people's health and promote well-being), 6 (water and sanitation), 11 (make cities inclusive, safe, resilient, and sustainable), and 13 (combating climate change and its impact) [17]. SDG-11 is closely related to solid waste management: "percentage of solid waste that is well-managed and collected on a regular basis." Yet, just like all prior social development programmes, operationalization and implementation may be where the problem lies. The management of solid waste is not mainstreamed, is funded poorly, and has never met expectations in many developing nations [27].

## 7. Conclusion

The mechanism of collection of solid waste and its safe disposal is a generic environmental problem worldwide. In this respect, the 3Rs-principle is considered to be the cheapest and most efficient way for solid waste management. In the majority of developing countries, there is no system for waste recovery at the collection and disposal sites. The recycling potential is not satisfactory. The practice of open burning of plastic and rubber items is the most common practice that causes air pollution. The waste management control mechanism is limited. Due to the lack of treatment facilities, people generally face unhygienic environments. Therefore, proper management must be done for solid waste management strategies.

## Abbreviation

| | |
|---|---|
| SW | solid waste |
| SWM | solid waste management |
| ISWM | integrated solid waste management |
| MSW | municipal solid waste |
| 3Rs | reduce, reuse, recycle |
| USA | United States of America |
| Cd | cadmium |
| Cr | chromium |
| Cu | copper |
| Ni | nickel |
| Pb | lead |
| SOx | oxides of sulpher |
| NOx | oxides of nitrogen |
| CO | carbon monooxide |
| $CH_4$ | methane |
| $H_2$ | hydrogen |
| SDGs | sustainable development goals |

## Author details

Nazish Huma Khan[1*], Nida Naz[1], Mohammad Nafees[2], Nida Gul[1] and Tooba Saeed[3]

1 Department of Environmental Sciences, University of Swabi, Pakistan

2 Department of Environmental Sciences, University of Peshawar, Pakistan

3 Department of Chemistry, Bacha Khan University Charsadda, Pakistan

*Address all correspondence to: humakhan@uoswabi.edu.pk

## References

[1] Memon MA. Integrated solid waste management based on the 3R approach. Journal of Material Cycles and Waste Management. 2010;**12**:30-40

[2] Arıkan E, Şimşit-Kalender ZT, Vayvay O. Solid waste disposal methodology selection using multi-criteria decision making methods and an application in Turkey. Journal of Cleaner Production. 2017;**142**:403-412

[3] Marshall RE, Farahbakhsh K. Systems approaches to integrated solid waste management in developing countries. Waste Management. 2013;**33**(4):988-1003

[4] Alam P, Ahmade K. Impact of solid waste on health and the environment. International Journal of Sustainable Development and Green Economics (IJSDGE). 2013;**2**(1):165-168

[5] Srivastava R, Krishna V, Sonkar I. Characterization and management of municipal solid waste: A case study of Varanasi city, India. International Journal of Current Research and Academic Review. 2014;**2**(8):10-16

[6] Soltani A, Sadiq R, Hewage K. Selecting sustainable waste-to-energy technologies for municipal solid waste treatment: A game theory approach for group decision-making. Journal of Cleaner Production. 2016;**113**:388-399

[7] Bao Z, Lu W. Developing efficient circularity for construction and demolition waste management in fast emerging economies: Lessons learned from Shenzhen, China. Science of the Total Environment. 2020;**724**:138264

[8] Bui TD, Tseng JW, Tseng ML, Lim MK. Opportunities and challenges for solid waste reuse and recycling in emerging economies: A hybrid analysis. Resources, Conservation and Recycling. 2022;**177**:105968

[9] Browning S, Beymer-Farris B, Seay JR. Addressing the challenges associated with plastic waste disposal and management in developing countries. Current Opinion in Chemical Engineering. 2021;**32**:100682

[10] Yousefloo A, Babazadeh R. Designing an integrated municipal solid waste management network: A case study. Journal of Cleaner Production. 2020;**244**:118824

[11] Khan NH, Nafees M, Saeed T, Rahman KU, Bashir A. Organizational setup for solid waste management record keeping. Journal of Engineering and Applied Sciences. 2021;**40**(1):60-68

[12] Menikpura SNM, Sang-Arun J, Bengtsson M. Assessment of environmental and economic performance of waste-to-energy facilities in Thai cities. Renewable Energy. 2016;**86**:576-584

[13] Shirazi M, Samieifard R, Abdoli M, Omidvar B. Mathematical modeling in municipal solid waste management: Case study of Tehran. Journal of Environmental Health Science & Engineering. 2016. DOI: 10.1186/s40201-016-0250-2

[14] Harijani A, Mansour S, Karimi B, Lee C. Multi-period sustainable and integrated recycling network for municipal solid waste. A case study in Tehran. Journal of Cleaner Production. 2017;**151**:96-108

[15] Louise Bjerkli C. Governance on the ground: A study of solid waste

Management in Addis Ababa, Ethiopia. International Journal of Urban and Regional Research. 2013;**37**(4):1273-1287

[16] Vinti G, Bauza V, Clasen T, Medlicott K, Tudor T, Zurbrügg C, et al. Municipal solid waste management and adverse health outcomes: A systematic review. International Journal of Environmental Research and Public Health. 2021;**18**(8):4331

[17] Ziraba AK, Haregu TN, Mberu B. A review and framework for understanding the potential impact of poor solid waste management on health in developing countries. Archives of Public Health. 2016;**74**(1):1-11

[18] Das S, Lee SH, Kumar P, Kim KH, Lee SS, Bhattacharya SS. Solid waste management: Scope and the challenge of sustainability. Journal of Cleaner Production. 2019;**228**:658-678

[19] Needhidasan S, Samuel M, Chidambaram R. Electronic waste–an emerging threat to the environment of urban India. Journal of Environmental Health Science and Engineering. 2014;**12**:1-9

[20] Nanda S, Berruti F. Municipal solid waste management and landfilling technologies: A review. Environmental Chemistry Letters. 2021;**19**:1433-1456

[21] Srivastava V, Ismail SA, Singh P, Singh RP. Urban solid waste management in the developing world with emphasis on India: Challenges and opportunities. Reviews in Environmental Science and Bio/Technology. 2015;**14**:317-337

[22] Singh RP, Singh P, Araujo AS, Ibrahim MH, Sulaiman O. Management of urban solid waste: Vermicomposting a sustainable option. Resources, Conservation and Recycling. 2011;**55**(7):719-729

[23] Khan NH, Nafees M, Saeed T, Khan A, Bashir A. Industrial Symbiosis and industrial waste Management in Wood-Based Industries. Journal of Industrial Pollution Control. 2018;**34**(2):2152-2158

[24] Sharif N, Pishvaee M, Aliahmadi A, Jabbarzadeh A. A bi-level programming approach to joint network design and pricing problem in the municipal solid waste management system: A case study. Resources, Conservation & Recycling. 2018;**131**:17-40

[25] Christensen T, editor. Solid Waste Technology and Management. Chichester: John Wiley & Sons Ltd; 2011. ISNB: 978-1-405-17517-3

[26] Venkiteela LK. Status and challenges of solid waste management in Tirupati city. Materials Today: Proceedings. 2020;**33**:470-474

[27] Haylamicheal ID, Desalegne SA. A review of legal framework applicable for the management of healthcare waste and current management practices in Ethiopia. Waste Management & Research. 2012;**30**(6):607-618

Chapter 3

# Perspective towards Construction and Demolition Waste Management: Case Study on Kahramanmaraş Earthquake

*Yasemin Tabak*

## Abstract

Türkiye is located in the most active earthquake zones with the shortest return period. There is plenty of demolition waste in 11 cities affected by the earthquake in Türkiye on 6th February 2023. A magnitude of 7.7 earthquake occurred in Pazarcik, Kahramanmaraş which is followed by an earthquake of 7.6 in Elbistan, Kahramanmaraş. This is the biggest disaster of the century. The number of collapsed or damaged buildings are approximately 280,000 and the economic damage is at least 100 billion US dollars. According to the Chamber of Environmental Engineers, 104 million tons of construction and demolition waste was generated. The demolition waste produced by the earthquake constitutes the solid waste with the highest weight and volume. The current earthquake rubble must be removed to attain the normal life in the cities. Thus it is necessary to develop and improve the technologies to be used for disposal and recycle of the waste stored in the areas under special conditions. In this book chapter, the evaluation of the wastes formed as a result of the 6th February 2023 earthquake, which is one of the largest earthquakes in the recent history of the world, will be discussed.

**Keywords:** Kahramanmaraş earthquake, construction, demolition, waste, disaster, debris

## 1. Introduction

Construction and demolition (C&D) waste originates from the demolition and renovation of concrete structures, roads, bridges and dams made of materials such as concrete, ceramics, sand, wood, brick, adhesive, metal, plastic. Construction and demolition waste comprises sand, soil, gravel, concrete, wood, bricks and masonry, plaster, metal, and asphalt. This waste can contain hazardous materials such as asbestos and lead. They are heavy, are often bulky, and occupy considerable storage space either on the road or in communal waste bin/container. The US Environmental Protection Agency (EPA) promotes a Sustainable Materials Management (SMM) approach that identifies particular construction and demolition materials as commodities that can be used in new building projects, thus avoiding the need to mine and process virgin materials. Some components of construction and demolition waste are precious as raw materials, while others are less valuable. However, they can still be easily converted into new products. Technology for the separating and recovering of construction and demolition waste is well established, easily accessible, and generally inexpensive. The global construction and demolition waste market is estimated as 26,622.1 million USD in 2021. It is projected to reach 34,407.0 million USD by 2026, at a compound annual growth rate (CAGR) of 5.3% during the forecast period. The global market is primarily driven by increasing construction activities and inclination of governments towards sustainability in various regions across the globe [1].

Construction and demolition wastes do not occur only as a result of building demolition, it is possible to say that the most significant amount of C&D waste is formed due to earthquakes. If we list the main earthquakes from 2000 to the present, it is possible to say that there are many significant earthquakes. On 26th December 2003, an earthquake measuring 6.3–6.6 on the Richter scale occurred in Bam, Iran. Peru was hit on 15th August 2007, by an earthquake of 7.9 on Richter scale [2]. On 12th May 2008, the Wenchuan earthquake or the Great Sichuan Earthquake happened in Beijing, which was a disastrous earthquake measured at 8.0 on the surface wave magnitude scale and 7.9 on the moment magnitude scale. During this earthquake, more than 90,000 people were killed or missing, the strongest earthquake in the last 50 years worldwide. A massive amount of building waste was generated by collapsed dwelling houses and dilapidated buildings after the Wenchuan earthquake which occurred on 12th May 2008, in Sichuan Province, PR China [3]. Haiti was hit on the evening of 12th January 2010, by an earthquake with a magnitude of 7.2 on the Richter scale [2]. Disasters have become more severe and frequent during the last few decades globally, and large disasters generate thousands of tons of waste, for example, the 8.9-magnitude earthquake and tsunami that hit Japan in 2011 generated 28 million tons of waste; likewise, Hurricane Katrina (2005) produced more than 100 million cubic yards of debris. This scale of waste presents secondary risks affecting the health of residents and causing environmental pollution [4]. In Central Italy, the 2016 earthquake devastated an area of almost 8000 $km^2$, affecting 140 municipalities and generating vast amounts of rubble. Hence, their C&D waste management is critical, undermining the reconstruction process of buildings [5].

Since it is a chapter on a case study, it is useful to give information about the major earthquakes in Türkiye. Türkiye is not subject to tornados or hurricanes, but earthquakes, landslides, floods, rock falls and avalanches frequently occur. The latter four disaster types are usually small-scale, with relatively little or no death toll. Earthquakes, however, are the most feared type of disasters in Türkiye, as many lives are often lost [6]. Türkiye is located on the seismically active Anatolian Plate. Considering past earthquakes, Türkiye has experienced major earthquakes. More than 20 earthquakes have occurred in this country since 1900, and Türkiye was among the first countries affected by earthquakes.

Between 1900 and 2023, Türkiye experienced 269 earthquakes that caused great economic damage and killed many people. Among these earthquakes, the first three earthquakes causing the most deaths and the most severe damage are; There were Kahramanmaraş Earthquake in 2023, Marmara Earthquake in 1999 and Erzincan Earthquake in 1939 [7]. In Türkiye since 1894, direct property and infrastructure losses arising from earthquakes have frequently exceeded 5 billion USD and, in the case of the 1939 Erzincan earthquake (7.9 Richter scale), have reached 23 billion USD [8].

Two devastating earthquakes occurred in Türkiye, one on 17th August 2009 and the other on November 12, 1999. The magnitudes were 7.4 and 7.2, respectively. The epicenter of the first earthquake was located near Gölcük, a town near Kocaeli province, 110 km from İstanbul. The epicenter of the second earthquake was in Düzce, 150 km from İstanbul [9]. Another earthquake with severe consequences for Türkiye was the 2011 Van earthquake. This earthquake, measured as 7.2 Richer scale caused destruction due to the existing state of the region's buildings, 644 people died and 1.966 people were injured [10].

In this book chapter, the evaluation of the wastes formed as a result of the 6th February 2023 earthquake, which is one of the largest earthquakes in the recent history of the world, will be discussed.

## 2. Effects of Kahramanmaraş earthquake

The catastrophic loss of life and damage caused by the two earthquakes that struck southern Türkiye and northern Syria on 6th February 2023 shocked the world. The devastating 7.8-magnitude earthquake near the Türkiye-Syria border in the early hours of 6th February 2023 was followed by another one nearly as strong as the first one. Thousands of people were killed and many more were injured. Thousands of buildings collapsed, leaving countless and homeless people exposed to unforgiving winter conditions. Schools and hospitals were destroyed [11, 12]. The earthquakes that occurred on 6th February 2023 and affected a total of eleven provinces, including Kahramanmaraş, Adıyaman, Hatay, Osmaniye, Gaziantep, Kilis, Şanlıurfa, Diyarbakır, Malatya, Adana and Elazığ in Eastern Türkiye were a natural disaster that caused significant damage to buildings and infrastructure, and claimed the lives of many people.

The earthquakes caused significant damage to buildings and infrastructure in almost all affected provinces. Many buildings were destroyed, with 8268 buildings in Hatay, 5000 in Adıyaman, 10,777 in Gaziantep, 338 in Şanlıurfa, 8633 in Kahramanmaraş, 434 in Diyarbakır, 1739 in Osmaniye, 33 in Adana, 5578 in Malatya and 447 in Kilis being either damaged or required demolition. As a result, a significant amount of construction and demolition waste was produced. The contents of this waste vary widely and include concrete, plaster, sand, gravel, wood, plastics, ceramics, metals, paper and cardboard, medical waste, and electronic waste. The recycling of these materials can contribute to the economy. Türkiye is one of the most earthquake-prone countries, and the places indicated in red on the map show the earthquake zones with high risk (**Figure 1**) [13].

Disaster waste management includes waste collection, transportation, reuse, recycling, and disposal during the emergency response, recovery, and reconstruction phases. Failure to manage disaster waste could result in prolonged recovery time, increased costs and potential risks to public health and the environment. Effective management of disaster waste could turn it into a valuable resource in the recovery and reconstruction process, thereby positively affecting social and economic recovery. Planning and coordination are essential for effective disaster waste management, which consists of three stages: waste requiring emergency response, recyclable waste, and reconstruction. These stages are not independent. The emergency response phase is considered a short-term action plan and involves the elimination of urgent threats to public health and safety, waste identification, characterization, and mapping. It typically lasts for a few days to two weeks. The recyclable waste phase is considered a

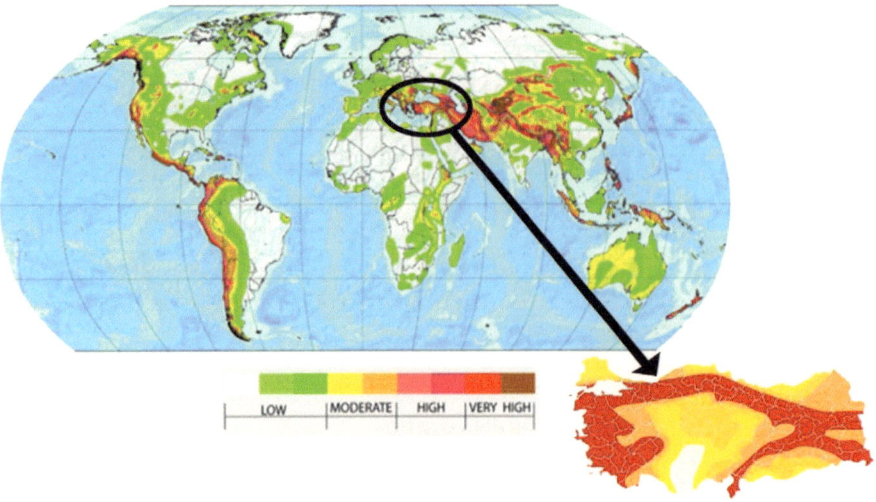


**Figure 1.**
*Representation of Türkiye earthquake zone.*

medium-term action plan, covering waste collection, transportation, and reuse/recycling logistics. It typically lasts for about five years. Finally, the reconstruction phase is considered a long-term action plan, covering continuous management projects design and evaluation of disaster waste situation. It typically lasts for about ten years.

Disaster waste management strategy is a discipline that involves waste planning and control, waste composition and quantity, waste collection, storage, transportation, processing, reuse, recycling, and disposal, based on environmental and public health, engineering, social conditions, legal frameworks, and financing.

The earthquakes also affected northern governorates in Syria, including Aleppo, Hama, Lattakia, Tartous and Idleb. These areas have been experiencing complex and protracted emergencies for nearly 12 years, with conflict and displacement complicated by recent outbreaks of cholera, measles, and COVID-19, on top of the already

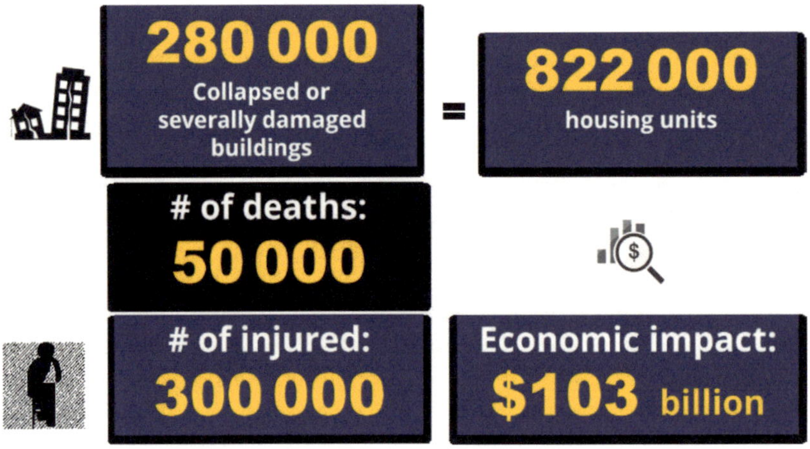


**Figure 2.**
*Impact assessment of Kahramanmaraş Earthquake.*

overwhelmed health system. These earthquakes are extraordinary both in terms of scale and in terms of immediate impact. During the winter season of the earthquake, ongoing aftershocks, lack of fuel for vehicles, damage to the power supply and communication infrastructure are still obstructing access and search and rescue efforts. In Türkiye, more than 4000 buildings have fallen and at least 15 hospitals have been affected with either partial or severe damage, causing the evacuation of its emergency rooms due to the risk of collapse. The runway of Hatay Airport was split and uplifted [14].

When the data collected as a result of the earthquake are evaluated, as summarized in **Figure 2**. It created 103 billion economic impacts, including 280,000 collapsed or severally damaged buildings and 822,000 housing units. In addition, there are over 50,000 deaths and over 300,000 injured people [13].

## 3. Construction and demolition wastes resulting from disasters

Estimating the composition of demolition waste generated by earthquakes provides a valuable lesson for the choosing of resource utilization pathways in post-disaster scenarios. It is clear that different building types yield varying amounts of demolition waste. As mentioned above, the buildings in Türkiye are mainly reinforced concrete and masonry structures, so the analysis is made for these two building types. Principal components and materials (such as walls, floors, stairs, windows, doors, etc.) were investigated before demolition, and material amounts of different components were recorded to estimate the proportion of various components of waste after demolition [15]. It is possible to say that 170 million tons of demolition waste (**Figure 3**) emerged due to the Kahramanmaraş earthquake [13].

Waste and debris management, collection, transportation, reuse, recycling or disposal of waste are included in the disaster management cycle's response/rescue/first aid and reconstruction/remediation stages [16]. Recycling and reuse are major components of disaster debris management with significant environmental, economic, and social benefits. Types of disaster debris and factors affecting management are given in **Table 1** [17, 18].

**Figure 3.**
*6th February 2023 earthquakes' demolition waste.*

| Waste category | Definition | Factors affecting management |
|---|---|---|
| Construction and demolition waste | The definition of C&D debris, which includes damaged or demolished buildings, roads, and other human-made structures such as lumber, plaster wall coverings, glass, metal, roofing materials, tiles, carpet, pipes, concrete, asphalt, auxiliary poles, wires, flooring, and furniture, may vary among waste management agencies at the state and local levels. | Waste materials generated from the demolition, renovation, or remodeling of buildings, known as construction and demolition (C&D) waste, can include materials that must be removed and used following federal standards, such as those containing asbestos insulation or tiles or transformers containing polychlorinated biphenyls (PCBs). C&D waste can also be mixed with materials that may affect whether the debris can be safely recycled and reused, or whether it must be burned, such as chemically treated or lead-based painted or wood products containing termites. |
| Domestic solid waste | Personal belongings and household garbage | It can be produced in volumes that exceed the existing landfill capacity or otherwise pollute non-hazardous waste |
| Vegetable waste | Fallen trees, branches, bushes and logs. | When they affect general access routes and critical infrastructure, they may need to removal immediately. They are often generated in large quantities that can be significantly reduced through incineration or shredding. Reduction and reuse options may be limited if they are contaminated. |
| Soil, mud and sand | Sediment is the soil material accumulating on property and rights of way due to landslides, intense winds or storm surges. | It can be reused as fill on residential or agricultural land. If it becomes contaminated with sewage, pesticide, fertilizer, or other unsafe chemicals for reuse, its options for reuse may be limited. |
| Putrefiable | Fruits and vegetables, meats, dairy products, and other products obtained from markets, restaurants, schools, hospitals, and residential areas that will decay. It may also include animal carcasses such as pets or farm animals. | It can be composted or processed to reduce volume. However, it should be collected and managed quickly to avoid attracting disease vectors such as rodents and flies. Decaying waste can pollute otherwise harmless waste streams if not managed quickly. |
| White goods | It is damaged or discarded household appliances such as refrigerators, freezers, air conditioners, washing machines, dryers, ovens, ranges, heat pumps, water heaters, and dishwashers. | The fridges might have been polluted by the rotting matter taken out. The proper management may be hard in case many white home goods are produced. |
| Vehicles and vehicles | Cars, trucks and boats damaged, destroyed or abandoned due to the incident. | If property and property issues are addressed, and hazardous liquids or materials (such as engine oil, gas and gas tanks, lead acid batteries, tires, airbags, and mercury switches) are emptied or removed, they can generally be recycled. |
| Household hazardous waste | Household products containing abrasive, toxic, flammable or reactive, engine oil, automobile batteries, paints and solvents, household cleaners and drain openers, components such as swimming pools, chemicals, pesticides and compressed gas tanks (such as propane and oxygen). | If hazardous waste from households is not collected and managed separately, the expense of waste management can be greater. Most jurisdictions prefer to separate these waste streams, but if they are blended together, it can pollute the ordinary household waste., |

| Waste category | Definition | Factors affecting management |
|---|---|---|
| Electronic waste (e-waste) | Computers, monitors, televisions, printers, stereos, DVD players and telephones. | Due to the presence of lead, chromium, cadmium, mercury, zinc and flame retardants, electronic waste is often separated for recycling by states. |
| Infected/ Medical waste | Waste that can cause infection in humans, such as contaminated animal waste, human blood and blood products, medical and pathological waste, and discarded sharps (needles, scalpels or broken medical instruments). | It can be mixed with and/or contaminated with harmless waste and pose a risk to unaware waste handlers. |
| Commercial or industrial hazardous waste | May contain petroleum or other hazardous substances (e.g. gas stations or dry cleaners) that pose a significant risk to human health, safety or the environment released from above-ground or underground storage facilities or containers (tanks or drums) or commercial or industrial facilities. | It may mix with and/or pollute with harmless waste. It can contaminate surface or groundwater if not quickly controlled, cleaned and managed correctly. |

**Table 1.**
*Types of disaster debris and factors affecting management.*

In the literature, wastes generated during disasters are generally examined in three stages [19–22].

i. Wastes that require urgent intervention,

ii. Recyclable waste,

iii. Not usable waste.

Removal of debris waste resulting from disasters is carried out in two stages. These stages are the first debris removal activity to eliminate life and security threats. The second is debris removal activities for the recovery of debris. Generally, the recovery phase of debris begins after emergency access roads have been cleared [16, 23]. Temporary storage areas are needed for collecting debris, classification and processing. These temporary areas must be predetermined for good debris management [24]. Temporary storage status of waste saves time for classification and disposal of waste. However, inappropriate land use may be limiting for this method. Temporary storage areas for planned activities such as storage, sorting and processing should be chosen considering ease of access, protection of environmentally sensitive areas and logistics efficiency [16, 22]. There are different disposal options for disaster waste. One disposal method for waste generated during disasters is the recovery of construction and demolition waste. In most cases, the main component of disaster waste is construction and demolition waste. Due to the lack of waste storage areas and limited natural resources, many research and development studies are carried out on the recovery of construction and demolition wastes [22, 25].

The wastes that occur as a result of the repair, modification, renewal, and demolition of housing, buildings, bridges, roads and similar infrastructure and superstructures or a natural disaster and whose components include materials such as concrete, rubble and steel are called "debris waste". Construction and demolition waste can be classified in several ways:

- According to their physical state: solid, liquid, gas, radioactive.
- According to its use: inert (inert; rubble, concrete), mobile (non-inert; frame, glass).
- According to hazards: dangerous (asbestos, PVC), non-hazardous (iron, gravel).

Although most construction and demolition wastes are in the category of harmless and inert wastes, some harmful or potentially harmful materials are used in structures. These materials (such as paint, adhesive, pitch) are immobilized in building production [26, 27].

Construction and demolition waste can be categorized as follows. These are:

i. Excavation materials: soil, sand, pebble, rock particles as well as the other materials forming through digging may be classified in this category. These kinds of wastes may also be formed by natural disasters such as floods and landslides. However, these materials' chemical structures depend on where the digging process takes place.

ii. Road construction and maintenance materials: these materials are asphalt, sand, gravel, metal, concrete.

iii. Debris wastes: These materials include soil, gravel, concrete chips, lime plaster, briquettes, veneer plates consisting of gypsum, sand, engineered stone and porcelain. Debris waste is not homogeneous. They are formed during the destruction of buildings and other structures.

iv. Work area waste materials: Such materials may be used for repair, support, augmentation, wood, plastic, paper, glass, metal, or rubber. They consist of paint, enamel, coating, glue and other materials [26, 27].

C&D waste is the solid waste produced by construction, demolishment and renovation, which constitutes almost 30%-40% of the global solid. The improper disposal of these wastes and their management can have negative effects on environment, economics and health of the human population. Most research on waste from C&D is confined to the reduction, recycling and reuse of waste.

It was impossible to rectify the damage to 147,895 buildings. (296,508 detached units). Allowing for current data, it is projected that debris removal must be implemented for a total of 35,355 buildings (96,100 detached units) as well as demolition and debris removal for 237,505 buildings (buildings requiring immediate demolition + severely damaged + moderately damaged (721,448 detached units)); thus, works on approximately 817,548 buildings in total are forecasted. By 6th March 2023, 94,297 businesses had been destroyed, urgently required demolition, or incurred serious harm. Predictions show that the total construction and debris waste will be between 100 and 120 million cubic meters [28]. Status of buildings collapsed or require demolition is given in the **Table 2**.

| Status of building number of buildings number of detached units | Status of building number of buildings number of detached units | Status of building number of buildings number of detached units |
|---|---|---|
| Collapsed | 34,355 | 96,100 |
| Requiring Urgent Demolition | 17,491 | 60,728 |
| Severely Damaged | 179,786 | 494,588 |
| Moderately Damaged | 40,228 | 166,132 |
| Not Yet Assessed for Damage | 147,895 | 296,508 |

**Table 2.**
*Status of buildings collapsed or requiring demolition.*

It will cost about 1.81 billion USD to transfer the construction and debris waste from the impacted areas to disposal locations. Regarding costs associated with the disposal plant, it is anticipated that stone crushing costs and construction and debris waste storage costs 406 million USD in total. These costs do not include the land cost of the facility. The overall cost of disposal for garbage from construction and debris in such case will be around 2.22 billion USD [28].

## 4. Management of construction and demolition waste

Waste management activities are designed according to the 4R golden rule given in **Figure 4**.

Conforming to the EU waste hierarchy, there are 3 facets of preventing waste concrete: cutting down the amount, reducing the damaging effect, and reducing the harmful content. Tactics to stop waste concrete consist of Eco-design, smart dismantling, and selective demolition, which are reported in **Table 3** [29].

The promotion of eco-design of buildings (detailed in **Table 2**) and the use of regulatory measures to enforce on-site dismantling, sorting, and selective demolition can be used to accomplish future C&D waste prevention. It is essential to separate C&D waste on-site to guarantee further reprocessing, as quality standards for waste intended for recycling or reuse can be very stringent in some cases. The amount of non-stony materials in the recycled concrete aggregate is limited to less than 1%, as non-stony residue such as glass would interfere with the alkali-silica reaction in new concrete products. On-site dismantling, sorting, and selective demolition must be used to separate contaminants before waste is recycled [29]. The wastes generated from construction, demolition and repair activities are transferred to storage areas and the environment. There are several ways they can be recovered without being sent back. It protects the environment and the economy. In construction and demolition less natural resources are used due to the evaluation of waste less energy consumed.

For the beneficial use of CW (construction waste) in building materials, primarily, It should be evaluated according to the "Waste Management Regulation-EK3B: Hazardous Waste Threshold Concentrations" published in the Official Gazette dated 02/04/2015 and numbered 29,314. Per the aforementioned regulation, whether the CW is dangerous or not should be determined by physical, chemical and ecotoxicological tests as well as organic and heavy metal analyses. In this way, it will also be

**Figure 4.**
*4R golden rule.*

| | Strategies at the design stage | Strategies at the EoL stage |
|---|---|---|
| Reduction of quantity | Long-lasting design, lightweight design, design for dismantling (DFD), design for deconstruction (D4D), design for recycling (DFR), designing out waste (DOW) | Smart dismantling, selective demolition |
| Reduction of adverse impact | DFF, D4D, DFR, DOW | Smart dismantling, selective demolition |
| Reduction of harmful content | DFF, D4D, DFR, DOW | Smart dismantling, selective demolition |

**Table 3.**
*Strategies for prevention of waste concrete.*

possible to determine waste codes (17 01 01, 17 01 02, 17 01 03, 17 02 01, 17 02 02, 17 02 03, 17 06 04, 17 08 02, 17 09 03) [30]. If CW is added in different proportions instead of silica sand as fine aggregate or gravel as coarse aggregate in concrete production, CW's geometric, physical and chemical properties must be tested according to TS 706 EN 12620+A1 [31].

Many construction materials that can be recovered from the CW can be recovered with the recovery units that can be permanently installed in the area where these wastes are transported for temporary storage. Reusing these recovered materials in appropriate areas will require both an environmentalist approach and compliance regarding the circular economy and zero waste practices. In addition, it is possible to examine and evaluate the environmental effects by handling the activities from the beginning to the completion of the process with Life Cycle Assessment (LCA) applications (**Figure 5**).

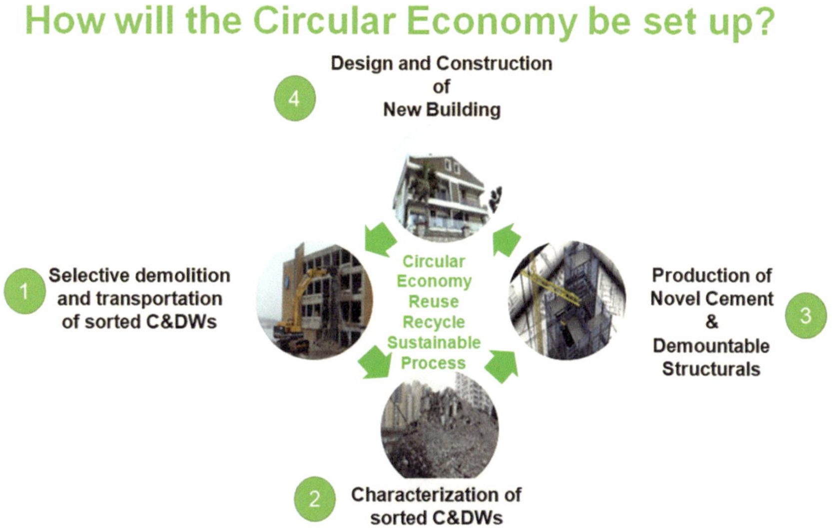


**Figure 5.**
*Life Cycle Assessment (LCA) applications [13].*

When it is desired to evaluate the situation by making LCA, after selective demolation and transportation of sorted C&D waste's, it may be possible to summarize the characterization of C&D waste as the production of bare cement and the production of disassembled structures, and finally the design and construction of new structures [13].

Considering the billions of tons of waste that is the common point of all Kahramanmaraş earthquake, recycling and reusing these wastes are essential. Concrete is the most widely used building material in the construction industry, mainly consisting of cement, sand, and aggregates. Therefore, using recycled concrete aggregates (RCAs) that emerged after C&D waste in concrete production is considered one of the best ways to overcome waste-related problems. Owing to the varying structural designs, levels of design, and construction management, the degree of damage and nature of destruction to buildings in the affected region varied greatly. Therefore, it is challenging to calculate the exact amount of building waste. The most reliable way is to make a sample survey of buildings of different architectural types in the disaster region.

One such solution is the recycling of waste concrete, which can be transformed into reusable material for new construction projects. The process involves crushing and grinding the waste concrete into small pieces, which are then sorted and cleaned to remove any contaminants. The resulting material, known as recycled concrete aggregate (RCA), can be used as a substitute for traditional aggregate in the production of new concrete. The use of RCA in concrete production offers several benefits, including reducing the demand for new concrete production, conserving natural resources, and minimizing waste generation. Recycling concrete can also reduce the amount of waste sent to landfills, which can help to alleviate environmental problems associated with landfill disposal. Moreover, using RCA in concrete production can lower the carbon footprint of construction projects, as it reduces the need for energy-intensive cement production.

There are possibilities to use selective or mixed construction and demolition wastes in concrete production, clinker production for the cement industry, and asphalt material in the brick-tile industry. Approximately optimum substitution utilization rates have been determined within the scope of some previous experimental studies and literature reviews. As known, substitution percentages may also vary depending on the fraction change of the waste [27].

20–45% of concrete wastes as aggregate in various concrete products (urban furniture, ready mixed concrete, lean concrete, screed concrete, concrete pipe, reinforced concrete pipe, paving stone, curbstone, manhole cover)

Up to 50% of concrete wastes as asphalt aggregate (road sub-base, plant mix base and bituminous base material)

100% of mixed construction debris waste in the road fill layer 3.9% of mixed construction debris wastes as raw material in clinker production

30% of the plastered brick and tile wastes into the brick raw material.

In addition, if asbestos can be separated from construction and demolition wastes; It is possible to save 20 billion TL (767.453.800,00 USD) in aggregate costs and 40 billion TL (1.534.907.600,00) in iron (**Figure 6**) [13].

While recycling waste concrete has been a common practice in many parts of the world for decades. However, with the growing interest in sustainable construction practices and the need for cost-effective solutions after the earthquake, now is an opportune time to explore the potential of recycling waste concrete in the country.

In Türkiye, the Ministry of Environment and Urbanization has developed policies and regulations for the management of construction and demolition waste. These policies aim to reduce the amount of waste sent to landfills and promote the use of recycled materials in construction.

In conclusion, the Kahramanmaraş earthquake presents an opportunity to rethink our approach to construction and waste management. By embracing innovative solutions like recycling waste concrete, we can reduce the environmental impact of reconstruction efforts and create a more sustainable future. Let us work together to build a better and more resilient world.

- The Turkish disaster response plan provides a rapid, effective, and comprehensive emergency response organization in the case of any possible disaster, saving more lives in a shorter period and in a wider area,

- By optimizing resources, response efforts will be executed rapidly.

**Figure 6.**
*Handling of 6th February 2023 earthquakes' demolition waste [13].*

- Economic and social losses will be minimized,

- The interruption of daily activities should be returned to normal in the shortest possible time. With the Disaster Recovery Plan, Türkiye is now more prepared for emergencies and disasters.

- Statistically, there are about 20 million housing in Türkiye. 14 million housing of this number have a disaster risk.

## 5. Conclusion

Türkiye is located in the most active earthquake zones with the shortest return period. There is plenty of demolition waste in 11 cities affected by the earthquake in Türkiye on 6th February 2023. A magnitude of 7.7 earthquake occurred in Pazarcik, Kahramanmaraş which is followed by an earthquake of 7.6 in Elbistan, Kahramanmaraş. This is the biggest disaster of the century.

Kahramanmaraş Earthquakes caused significant damage to buildings and infrastructure in almost all affected provinces. Many buildings were destroyed, with 8268 buildings in Hatay, 5000 in Adıyaman, 10,777 in Gaziantep, 338 in Şanlıurfa, 8633 in Kahramanmaraş, 434 in Diyarbakır, 1739 in Osmaniye, 33 in Adana, 5578 in Malatya and 447 in Kilis being either damaged or required demolition. As a result, a significant amount of construction and demolition waste was produced. The contents of this waste vary widely and include concrete, plaster, sand, gravel, wood, plastics, ceramics, metals, paper and cardboard, medical waste, and electronic waste. The recycling of these materials can contribute to the economy.

After the earthquake relief, the disposal of large amounts of demolition waste become an urgent problem for the governments. Improper disposal of demolition waste will further deteriorate the environment and cause resource waste.

Construction and demolition (C&D) waste originates from the demolition and renovation of concrete structures, roads, bridges and dams made of materials such as concrete, ceramics, sand, wood, brick, adhesive, metal, plastic. Construction and demolition waste comprises sand, soil, gravel, concrete, wood, bricks and masonry, plaster, metal, and asphalt. This waste can contain hazardous materials such as asbestos and lead.

Some components of construction and demolition waste are precious as raw materials, while others are less valuable. However, they can still be easily converted into new products. Technology for the separating and recovering of construction and demolition waste is well established, easily accessible, and generally inexpensive. The global construction and demolition waste market is estimated as 26,622.1 million USD in 2021. It is projected to reach 34,407.0 million USD by 2026, at a compound annual growth rate (CAGR) of 5.3% during the forecast period.

Considering the billions of tons of waste that is the common point of all Kahramanmaraş earthquake, recycling and reusing these wastes are essential. Concrete is the most widely used building material in the construction industry, mainly consisting of cement, sand, and aggregates. Therefore, using recycled concrete aggregates (RCAs) that emerged after C&D waste in concrete production is considered one of the best ways to overcome waste-related problems.

Owing to the varying structural designs, levels of design, and construction management, the degree of damage and nature of destruction to buildings in the affected region varied greatly. Therefore, it is challenging to calculate the exact amount of building waste. The most reliable way is to make a sample survey of buildings of different architectural types in the disaster region.

In Türkiye, the Ministry of Environment and Urbanization has developed policies and regulations for the management of construction and demolition waste. These policies aim to reduce the amount of waste sent to landfills and promote the use of recycled materials in construction.

One such solution is the recycling of waste concrete, which can be transformed into reusable material for new construction projects. The process involves crushing and grinding the waste concrete into small pieces, which are then sorted and cleaned to remove any contaminants. The resulting material, known as recycled concrete aggregate (RCA), can be used as a substitute for traditional aggregate in the production of new concrete. The use of RCA in concrete production offers several benefits, including reducing the demand for new concrete production, conserving natural resources, and minimizing waste generation. Moreover, using RCA in concrete production can lower the carbon footprint of construction projects, as it reduces the need for energy-intensive cement production.

In conclusion, the Kahramanmaraş earthquake presents an opportunity to rethink our approach to construction and waste management. By embracing innovative solutions like recycling waste concrete, we can reduce the environmental impact of reconstruction efforts and create a more sustainable future. Let us work together to build a better and more resilient world.

- The Turkish disaster response plan provides a rapid, effective, and comprehensive emergency response organization in the case of any possible disaster, saving more lives in a shorter period and in a wider area,

- By optimizing resources, response efforts will be executed rapidly.

- Economic and social losses will be minimized,

- The interruption of daily activities should be returned to normal in the shortest possible time. With the Disaster Recovery Plan, Türkiye is now more prepared for emergencies and disasters.

- Statistically, there are about 20 million housing in Türkiye. 14 million housing of this number have a disaster risk.

## Conflict of interest

The author declares no conflict of interest.

## Notes/thanks/other declarations

I would like to thank Dr. Tülay Çağlayan Özlü, Assoc. Prof. Şeyda Polat and MSc. Beril Pınar Özler. Kristina Kardum Cvitan, Author Service Manager(s) at IntechOpen for their constant support, communication and feedback.

## Author details

Yasemin Tabak
TUBITAK Marmara Research Center, Materials Technologies, Gebze, Kocaeli, Türkiye

*Address all correspondence to: yasemin.tabak@tubitak.gov.tr;
dryasemintabak@gmail.com

## References

[1] Construction & Demolation Waste Market. Global Forecast to 2026, By Type (Sand, Soil & Gravel, Concrete, Bricks & Masonry, Wood, Metals & Others), Source (Residential, Commercial, Industrial, Municipal) and Region, MarketsandMarkets. 2021

[2] Motamedi MHK, Sagafinia M, Ebrahimi A, Shams E, Motamedi MK. Major earthquakes of the past decade (2000-2010): A comparative review of various aspects of management. Trauma Monthly. 2012;**17**(1):219-229. DOI: 10.5812/traumamon.4519

[3] Xiaoa J, Xiea H, Zhang C. Investigation on building waste and reclaim in Wenchuan earthquake disaster area. Resources Conservation and Recycling. 2012;**61**(04):109-117. DOI: 10.1016/j.resconrec

[4] Lee Y, Kim Y, Lee H. Framework for selection of temporary disaster waste management sites for post-flood recovery in Seoul, South Korea. International Journal of Disaster Risk Reduction. 2022;**71**:102832. DOI: 10.1016/j.ijdrr

[5] Volpintesta F, Ossoli E, Reggiani A, Stabile P, Santulli C, Paris E. Geopolymers-based application for the up-cycling utilization of construction and demolition waste from the 2016 central Italy earthquakes. Materials Letters. ;**2023**(336):133849. DOI: 10.1016/j.matlet

[6] Kılcı F, Kara BY, Bozkaya B. Locating temporary shelter areas after an earthquake: A case for Turkey. European Journal of Operational Research. 2015;**243**:323-333. DOI: 10.1016/j.ejor.2014.11.035

[7] Government of Türkiye. Türkiye Earthquakes Recovery and Reconstruction Assessment [Assessment]. The European Union Delegation in Türkiye and the EU Joint Research Centre; 2023

[8] Gurenko E, Lester R, Mahul O, Gonulal SO. Earthquake Insurance in Turkey. History of the Turkish Catastrophe Insurance Pool. The World Bank; 2006 ISBN-13: 978-0-8213-6583-0 ISBN-10: 0-8213-6583-5

[9] Sahin M, Tari E. The August 17 Kocaeli and the November 12 Duzce earthquakes in Turkey. Earth, Planets and Space. 2000;**52**:753-757. DOI: 10.1186/BF03352277

[10] Uzun A, Oğlakci B. Turkey's Earthquake history and institution based earthquake reduction policies and strategies. Earthquake Rısk Perception, Communıcatıon and Mıtıgatıon Strategıes across Europe. 2019;**2**(2)

[11] Çetin KÖ, Ilgaç M, Can G, Çakır E. Middle East Technical University, Preliminary Reconnaissance Report on February 6, 2023, Pazarcık Mw=7.7 and Elbistan Mw=7.6, Kahramanmaraş-Türkiye Earthquakes. Report No: METU/EERC 2023-01. 2023

[12] Mavrouli M, Mavroulis S, Lekkas E, Tsakris A. An emerging health crisis in Turkey and Syria after the earthquake disaster on 6 February 2023: Risk factors, prevention and Management of Infectious Diseases. Healthcare. 2023;**11**(7):1022. DOI: 10.3390/healthcare11071022

[13] ICEBERG Project. Innovative Circular Economy Based Solutions Demonstrating the Efficient Recovery of Valuable Material Resources from the Generation of Representative End-of-Life

Building Materials, H2020-EU.3.5.3 (GA n 869336). 2020-2024. https://iceberg-project.eu/

[14] World Health Organization. WHO flash appeal Earthquake response in Türkiye and Whole of Syria [Internet]. 2023. https://www.who.int/publications/m/item/who-flash-appeal--earthquake-response-in-t-rkiye-and-whole-of-syria [Accessed: May 30, 2023]

[15] Xiao J, Deng Q, Hou M, Shen J, Gencel O. Where are demolition wastes going: Refection and analysis of the February 6, 2023 earthquake disaster in Turkey. Low-carbon Materials and Green Construction. 2023;**1**:17. DOI: 10.1007/s44242-023-00017-3

[16] Karunasena G, Amaratunga D, Haigh R, Lill I. Post disaster waste management strategies in developing countries: Case of Sri Lanka. International Journal of Strategic Property Management. 2009;**13**(2):171-190. DOI: 10.3846/1648-715X.2009.13

[17] FEMA. Public Assistance Debris Management Guide FEMA-325, Appendix D: Debris Management Plan Job Aid. 2017. https://www.fema.gov/media-library/assets/documents/111781 [Accessed: May 30, 2023]

[18] EPA, Guidance Planning for Disaster Debris. 2008. https://www.epa.gov/homeland-security-waste/guidance-aboutplanning-natural-disaster-debris [Accessed: May 30, 2023]

[19] Kuramoto N. The actual state of damage and measures under taken in Hyogo prefecture. In: Earthquake Waste Symposium,12-13 June 1995. Osaka; 1995

[20] Baycan F, Petersen M. Disaster waste management- C&D waste. In: Annual Conference of the International Solid Waste Association, 8-12 July 2022. Istanbul, Turkey; 2022

[21] JEU. Joint Environmental Unit Disaster Waste Management Guidelines, Final Draft [Internet]. 2010. https://www.humanitarianlibrary.org/sites/default/files/2014/02/JEUDWMPFinalDraftfordistribution.pdf [Accessed: July 12, 2018]

[22] Güler ÜA. Sürdürülebilir Afet Yönetiminde Atık Yönetimi, Derleme/Review. Doğ Afet Çev Derg. 2018;**4**(2):236-246. DOI: 10.21324/dacd.412380

[23] FEMA. Public Assistance Debris Management Guide. Washington: Federal Emergency Management Agency (FEMA); 2007 https://www.fema.gov/media-library-data/20130726-1826-25045-418/fema_325__debris_management_guide_2007.07.25.pdf, [Accessed: July 12, 2018

[24] EPA. Planning for Disaster Debris. United States Environmental Protection Agency (EPA); 2008 https://www.epa.gov/sites/production/files/2016-06/documents/20160629-16-p-0219.pdf [Accessed: July 12, 2018

[25] Lauritzen EK. Disaster Waste Management. Construction, Demolition and Disaster Waste Management. 1st ed. CRC Press; 2018. p. 71

[26] Günay E, Kara M, Öztürk Y, Kılıç Y, Sarıgüzel M, Tabak Y. Evoluation of Construction and Demolation Waste Project Report. TUBITAK; 2012

[27] Kara M, Şahin A, Kılıç Y. Construction and Demolition Waste Recycling and Determination of Usage Criteria Project Report. TUBITAK; 2013

[28] Lorca A, Celik M, Ergun Ö, Keskinocak P. An optimization-based decision-support tool for post-disaster

debris operations. Production and Operations Management. 2016;**26**(6):1076-1091. DOI: 10.1111/poms.12643

[29] Zhang C, Hu M, Maio F, Sprecher B, Yang X, Tukker A. An overview of the waste hierarchy framework for analyzing the circularity in construction and demolition waste management in Europe. Science of the Total Environment. 2022;**803**:149892

[30] General Directorate of Environmental Management of Türkiye. Waste Management Regulation-EK3B: Hazardous Waste Threshold Concentrations. 2015. p. 29314

[31] Turkish Standards Institute, TS 706 EN 12620 +A1:2009 Concrete Aggregates Standard

Chapter 4

# An Investigation of Waste Management Practice in a South African Township: A Case Study of Ekuphumleni Township, Ndlambe Municipality

*Ayo Adeniran, Lorato Motsatsi, Sijekula Mbanga, Emma Ayesu-Koranteng and Winston Shakantu*


## Abstract

Solid waste is a global challenge that is more pronounced in developing countries such as South Africa, where its management is a major concern. The government has recently made a concerted effort to engage the public in sustainable waste management practices to resolve the crisis occasioned by the challenge. This chapter investigates waste management practices in the South African Township of Ekuphumleni and relied on a sample of 353 households to obtain some primary data with a questionnaire on the subject matter. The data collected was analyzed using "R," and the results were presented using charts, tables, and figures. Data collected revealed that waste paper, cans, used plastics, and bottles were major waste components generated by the respondents and these wastes were generally stored unseparated domestically in plastic bags and home garbage can. Furthermore, the respondent indicated that the municipality does a door-to-door collection of their waste and they were unwilling to pay for waste collection services. While the waste management practice is in tandem with the municipal system, the study recommends that the respondents must be educated on circularity, which will ensure reducing, reusing, recycling, and recovering waste and further aid economic empowerment.


**Keywords:** waste, waste management, township, South African, waste separation, waste storage

## 1. Introduction

Waste continues to be generated because of human activities, and as the increase in the human population stimulates urbanisation, it is becoming an issue of global concern [1]. Waste refers to all pieces of objects and items such as garden waste, packing items, vegetables, metals and old paint containers, among others, that owners have no more use for and they aim to discard [2, 3]. The concern for such items stems from the continuous contamination of the atmosphere, soil and water, which endlessly impacts public health and global degradation [4].

It is important to note that the increasing solid waste generation places additional strain on the already overburdened waste management systems, and if waste is not properly managed, it may cause societal problems with the "Not-In-My-Backyard" mindset anticipated to emerge and prevail [5]. Besides, poorly managed solid wastes can have catastrophic environmental implications, such as becoming a breeding ground for disease-spreading vectors, production of leachates which contaminates groundwater, production of methane gas with its subsequent effects on global warming and climate change and increased fire outbreak, to mention a few [6].

Waste management is simply the collection, transportation, processing, or disposal of waste materials [7]. Chand [8] further described waste management as a procedure to mitigate the waste impact on the environment, health, or aesthetics. However, the poor handling of the procedure in urban and rural areas has been a major problem for human health and existence [9].

As a result of the global impact of waste, at least 12 of the 17 Sustainable Development Goals (SDGs) of the 2030 Agenda for Sustainable Development, adopted by the 193 UN Member States in September 2015 [10], have a direct association with solid waste management. Furthermore, according to the Global Waste Management Outlook (GWMO), the cross-cutting nature of solid waste management (SWM) and its impact on 12 SDGs emphasise its importance and political priority [11].

Despite the SDG's focus, waste and its management practices remain a major global challenge [12]. Low-income countries' main waste procedures and disposal mechanisms are open dumping and open burning [13]. For example, open littering [14], open dump [15], illegal dump sites [16], and incineration [17], among others, are some of the waste management practices still in practice in developing countries.

While no study as to the practice of waste management in Ekuphumleni Township, Ndlambe Municipality has been conducted, this chapter seeks to present the findings of the current practice and make recommendations towards addressing and raising the level of awareness and knowledge of sustainable solid waste management practices in the low-income neighbourhood of emerging nations.

The subsequent sections present the literature works, the methodology employed, findings and recommendations and conclusions of the study.

## 2. Waste management practices

Waste is unwanted, useless, and discarded material generated daily by human activities [18]. The E.U. Waste Directive defines waste as any object or substance the owner throws away, implying that it is useless [19]. However, several schools of thought, such as Steenmans and Malcolm [20], Thomas [21], and Hannon and Zaman [22], have argued against this definition as the value of waste is deemed to be subjective as what is waste to a consumer is a resource to another.

As a result, Wiprächtiger [23] argues that there is no such element as final waste because its definition will always depend on the degree of its perceived usefulness to its holder. It is then safe to align with van Ewijk [24] that the definition of waste is always contextual and can depend on the prevailing state of technology, the environment, and political ideology.

According to the Basel Convention, waste is categorised into two main typologies: hazardous and non-hazardous [25]. While hazardous waste is regulated at the national level, regulating the non-hazardous is within the purview of the regional and or municipal government [26]. The Basel convention documented hazardous wastes as radioactive, industrial, electronic and medical waste, among others, while the non-hazardous are municipal and non-hazardous industrial waste [27].

Mngomezulu [14] and Adeniran [15] identified the typology of municipal waste to include cans, and other metals, paper, bottles, plastics, food remains, old appliances, glass and construction demolition waste, among others.

Hoornweg and Bhada-Tata [6] further identified the types of waste and their sources to be: glass (broken glassware and bottles, coloured glass and light bulbs, among others); metal (foil, cans, tins, appliances and railings, among others); organic (garden/yard waste, food scraps and wood process residues, among others); paper (newspaper, cardboard, paper scraps and boxes, among others); plastic (packaging, containers, bags and lids, among others); and other (leather, textiles, rubber, multilaminates and other inert materials).

The growth in waste is alarming in metropolitan areas, and this is due to population movements towards these centres [28]. Waste growth tends to rise proportionately with urbanisation, rising income levels, and population expansion [12, 29]. While the global population keeps growing, its changing demographics are quickly evolving, and such areas are witnessing unprecedented levels of urbanisation, with the majority of this growth occurring in small and medium-sized cities in low-income countries [30]. Amaral [31] indicates that the unprecedented population growth has several environmental consequences, including increased urbanisation and municipal solid waste generation, which is expected to reach 3.4 billion tonnes annually by 2050. This waste growth is unfortunately not being matched with appropriate management practices [32].

Waste management is collecting, storing, treating, and disposing of waste materials in a manner that is safe for humans, vegetation, living creatures, ecosystems, and the environment [33].

As practiced in most emerging countries, households dispose of all forms of waste together [34], and the municipal trucks collect them [35] and when and if not collected, the practice of illegal dumping, littering and open burning of waste is practiced [36].

In South Africa, AWARD [37] indicated that over 90% of the collected waste is disposed of into landfills.

The literature highlighted three types of landfills: the open dump, the semi-controlled landfill, and the sanitary landfill [38]. Despite the attendant challenges posed by the open dump practices, it is still the most used method by urban centres in the developing world [12].

Waste collection and landfill activities have significantly contributed to greenhouse gas (GHG) emissions and climate change [39]. However, these poor waste management practices have fallen excessively on the poverty-stricken neighbourhoods with little or no influence on the waste products being illegally dumped near them [40].

A waste management system includes appropriate separation and decommissioning, logistics, storage, worker training and disposal facilities [41].

Adeniran [42] posited that numerous policies had been positioned to tackle waste and pollution in Sub-Saharan Africa. However, [43] argues that it is unclear if these policies are actioned as there is little or no progress towards achieving their aims and obligations. In addition, UNEP [44] indicated that the inability of many African governments to enforce waste and environmental regulations had fostered an environment of impunity, thus affecting the performance of waste management. According to David [45], the resultant effect is that industry participants are incapable of keeping up with the increasing waste streams and the timely development of strategies and policies to manage them effectively.

Despite their limited capacity for planning, limited resources, operational monitoring, and contract management, local governments are frequently in charge of an effective waste management system, and these limiting factors make sustainable waste management difficult [46].

Globally, various waste handling and disposal systems are in place; however, the major difference between the systems of advanced and emerging economies is waste separation at the collection point [47], which facilitates waste recycling and reuse, recognised as the most beneficial waste management system.

Mir [48] aver that the population must accept a waste management system to be effective, and [49] underscores the importance of ensuring a higher standard of living for future generations, simply defined as sustainability. Hence a solid waste management system must be socially acceptable, economically viable, and environmentally efficient to be sustainable [50].

Affordability denotes that all sectors of society accept the cost of maintaining a clean environment, whereas societal acceptance denotes that the inhabitants agree to the service offered if it meets their needs [51]. Meanwhile, the waste management system must be environmentally friendly by implementing an environmental conservation strategy, structure, and policy.

Simatele [52] documented that South Africa, like other developing countries, has implemented waste management policies, but their application proved inconsistent. Dlamini *et al.* indicated that these policies cover a set of efforts to tackle enhancements for environmental and public health quality. Nonetheless, despite the legal importance and quality, the law's enactment per se does not guarantee improvements in solid waste management [53].

South Africa has 13 pieces of legislation on waste management [54]; the most recent is the National Environmental Management: Waste Amendment Act, 2014 (Act 26 of 2014). The thrust of this legislation is *"to protect health and the environment by providing reasonable measures for the prevention of pollution and ecological degradation and for securing ecologically sustainable development".*

## 3. Methodology

The data for this paper were collected between 20th and 24th September 2020 between the hours of 10 h00 and 20 h00 to ensure that respondents who had gone to work were given the opportunity as they are expected to have returned by the evening at the latest. The time frame was chosen because residents are expected to have completed their household chores by 10 h00; those who had left home in the morning would have returned for dinner by the late evening. During the collection period, data were collected on various aspects, including household waste management and disposal practices; thus, this study focuses on the waste management practices of Ekuphumleni households. Ekuphumleni township is adjacent to the Kenton on Sea within the Ndlambe Municipality in the Eastern Cape, South Africa. The township is located approximately 130 km from Port Elizabeth on the Port Alfred-East London corridor.

| Mean | Decision/Interpretation | | |
|---|---|---|---|
| **1.00 to 1.44** | **Never** | **Very difficult** | **Very Unsafe** |
| 1.45 to 2.44 | Rarely | Difficult | Unsafe |
| 2.45 to 3.44 | Sometimes | Moderate | Fairly safe |
| 3.45 to 4.44 | Often | Easy | Very safe |
| 4.45 to 5.00 | Always | Very easy | |

*Adapted from Sarrafzadeh [56].*

**Table 1.**
*Decision rule.*

The estimated population of the township was about 1800 households and using a 95% confidence level and a 5% margin of error, 317 households were targeted, but we succeeded in reaching a sample of 353 households using convenience sampling to gather primary data from the willing and available representatives of all households during the fieldwork.

Within the COVID-19 protocol, this study utilised a Likert scale-like questionnaire; because of its simplicity in composition, the Likert Scale was the preferred scaling system for applicable statements/questions as it also allows for the use of hidden perceptions and is expected to yield a high accuracy of measurement [55]. The questionnaire covered a wide range of topics and offered information for developing a local economic strategy as the data collection tool. The data collection was managed by the researchers, who also participated in the data collection, assisted by a team of well-trained field workers. There had been several meetings and consultations between all stakeholders regarding the green village project to be developed within the community prior to administering the questionnaire. Ward Councillors, Community Representatives, and Municipality Officials attended these meetings and expressed their support for conducting the study in the area. After approval, potential participants were approached, informed consent was obtained from them, all the participants were assured of anonymity and confidentiality, and their participation was entirely voluntary.

The data collected was analysed using SPSS, and the adopted decision rule was adapted from [56] and presented in **Table 1**.

## 4. Findings

### 4.1 Demography

Demographic data allows us to determine whether there are differences in the answers provided by the respondents based on personal characteristics, and it also assists us in determining if there are gaps in our data, allowing us to ensure that it reflects the subject in question [57].

#### *4.1.1 The population of households by gender*

The data collected on the gender makeup of each household is presented in **Figure 1**, and it shows that the 353 participating households have a female population of 748 (57.2%) and a male population of 560 (42.8%). This implies that each

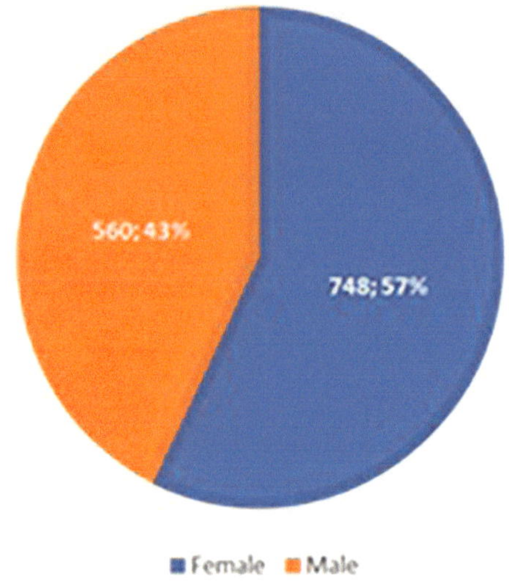


**Figure 1.**
*Population of household by gender.*

household has an average of 3.70 people, i.e. the ratio of females to males is 2.12 to 1.58 per household).

### *4.1.2 Household headship*

The questionnaire did not specify how participants should perceive headship, and no question queried what made a household member the head. However, in most households, the person described as the head of the household was the oldest family member.

As shown in **Figure 2**, except for one household with a coloured male respondent head, there are more black female-headed households than black male-headed households across all age groups. According to the frequency distribution in **Figure 3**, the overall mean age of the household head was 46 years.

### *4.1.3 Education level*

Using households that are 20 years and above, **Table 2** shows that 410 respondents representing 93.0% of the household members either did not attend school, had incomplete or complete primary and incomplete and complete secondary school while 17 Nr (3.9%) have certificate and 10 Nr (2.3%) hold diplomas. It is also interesting to note that the 4 Nr (0.9%) with a bachelor's degree are all female. Again, these figures apply to household members (20 years and above) whose highest education qualification was reported.

### *4.1.4 Household average monthly income*

As revealed in **Tables 3**, 244 households (69.1%) live on an average monthly income of less than R6000, while another 46 households (13%) earn no income.

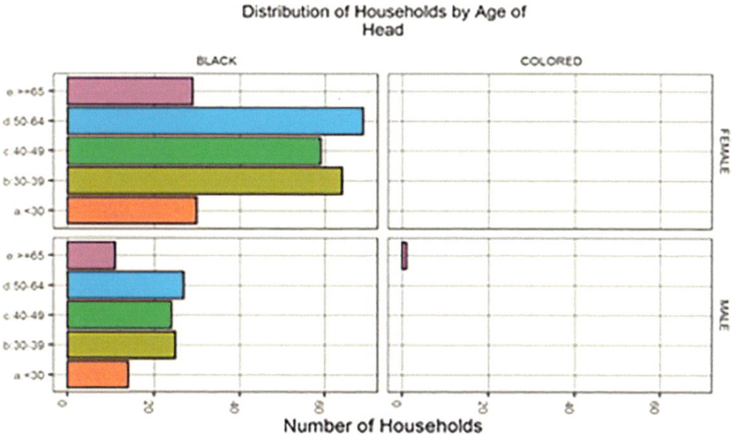


**Figure 2.**
*Age of household head.*

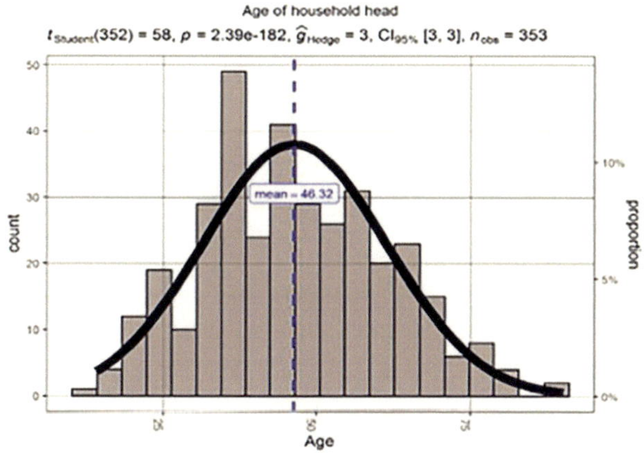


**Figure 3.**
*Distribution of age of household head.*

6.2% earn an average monthly income of between R6000 and above R20000, meaning they are in the mid to high-income strata.

## 4.2 Waste management practices in Ekuphumleni township

### *4.2.1 Types of waste generated and frequency*

Taking a cue from **Table 1**, as seen from **Table 4**, waste paper, cans, used plastics, and bottles rank first, second, third and fourth with mean scores of 3.13, 3.00, 2.95, and 2.92, indicating that the respondents sometimes generate these materials as waste. On the other hand, food remains, old clothing, old appliances, human waste, hazardous waste and oil are rarely generated as their mean scores ranged between 1.45 and 2.44; decision rule from **Table 1**, the respondents never generate other types of waste with a mean score of 0.32. With a composite mean of 2.09, there is a sign that waste is rarely generated in the township.

| Response | Female | | Male | | Total | |
|---|---|---|---|---|---|---|
| | Nr | % | Nr | % | Nr | % |
| No schooling | 30 | 6.8 | 25 | 5.7 | 55 | 12.5 |
| Incomplete primary | 59 | 13.4 | 44 | 10.0 | 103 | 23.4 |
| Complete primary | 12 | 2.7 | 17 | 3.9 | 29 | 6.6 |
| Incomplete Secondary | 72 | 16.3 | 59 | 13.4 | 131 | 29.7 |
| Complete secondary | 54 | 12.2 | 38 | 8.6 | 92 | 20.9 |
| Certificate | 13 | 2.9 | 4 | 0.9 | 17 | 3.9 |
| Diploma | 5 | 1.1 | 5 | 1.1 | 10 | 2.3 |
| Bachelors' degree | 4 | 0.9 | 0 | 0.0 | 4 | 0.9 |
| Total | 249 | 56.5 | 192 | 43.5 | 441 | 100 |

**Table 2.**
*Household members' highest educational qualification.*

| Income band (Rands) | Number of households | Per cent |
|---|---|---|
| 1 to 1999 | 114 | 32.3 |
| 2000 to 3999 | 108 | 30.6 |
| 4000 to 5999 | 22 | 6.2 |
| 6000 to 9999 | 9 | 2.5 |
| 10,000 to 12,999 | 5 | 1.4 |
| 13,000 to 15,999 | 6 | 1.7 |
| 16,000 to 20,000 | 1 | 0.3 |
| >20,000 | 1 | 0.3 |
| Unspecified | 4 | 1.1 |
| None | 46 | 13.0 |
| | 353 | 100 |

**Table 3.**
*Household average monthly income.*

### *4.2.2 Waste storage material and frequency*

Plastic bags with a mean score of 3.45 is often used as storage material by the respondents, as shown in **Table 5**, and it ranks first. Home garbage cans (MS, 2.11) and cardboard boxes (MS 1.68), ranking second and third respectively, are often used, while municipal plastic drums (MS 1.43), biodegradable sacks (MS 1.20), nearby municipal dumpster (MS 0.92) and into unused open plots (MS 0.75) are never used. With a composite mean of 1.65, there is an indication of general apathy towards storing waste in materials.

### *4.2.3 Waste separation*

**Table 6** reveals that 323 respondents representing 91.5%, indicated that they do not separate their wastes, while 19 Nr (5.4%) stated that they do and 11 Nr (3.1%)

| Types of waste | Response | | | | | | | | |
|---|---|---|---|---|---|---|---|---|---|
| | A | O | S | R | N | U | Total | MS | R* |
| Wastepaper | 110 | 70 | 56 | 29 | 49 | 39 | 353 | 3.13 | 1 |
| Cans | 91 | 79 | 59 | 29 | 53 | 42 | 353 | 3.00 | 2 |
| Used plastics | 74 | 85 | 73 | 35 | 45 | 41 | 353 | 2.95 | 3 |
| Bottles | 71 | 77 | 83 | 40 | 40 | 42 | 353 | 2.92 | 4 |
| Food remains | 38 | 23 | 92 | 59 | 101 | 40 | 353 | 2.20 | 5 |
| Old clothing | 20 | 37 | 71 | 67 | 111 | 47 | 353 | 2.00 | 6 |
| Old appliances | 14 | 15 | 59 | 130 | 91 | 44 | 353 | 1.86 | 7 |
| Human waste | 45 | 20 | 26 | 22 | 148 | 92 | 353 | 1.63 | 8 |
| Hazardous waste | 14 | 11 | 50 | 55 | 149 | 74 | 353 | 1.48 | 9 |
| Oil | 22 | 11 | 25 | 40 | 208 | 47 | 353 | 1.46 | 10 |
| Others | 7 | 4 | 10 | 3 | 26 | 303 | 353 | 0.32 | 11 |
| Composite Mean Score = 2.09 | | | | | | | | | |

*Key: A = Always; O = Often; S = Sometimes; R = Rarely; N = Never; U = Unspecified; MS = Mean Score; and R* = Ranking.*

**Table 4.**
*Types of waste generated and frequency.*

| Waste storage material | Response | | | | | | | | |
|---|---|---|---|---|---|---|---|---|---|
| | A | O | S | R | N | U | Total | MS | R* |
| Plastic bags | 200 | 32 | 20 | 8 | 12 | 81 | 353 | 3.45 | 1 |
| Home garbage can | 110 | 17 | 14 | 17 | 52 | 143 | 353 | 2.11 | 2 |
| Cardboard boxes | 45 | 21 | 55 | 26 | 68 | 138 | 353 | 1.68 | 3 |
| Municipal plastic drums | 49 | 20 | 14 | 27 | 85 | 158 | 353 | 1.43 | 4 |
| Biodegradable sacks | 27 | 16 | 24 | 27 | 100 | 159 | 353 | 1.20 | 5 |
| Nearby municipal dumpster | 8 | 12 | 22 | 35 | 100 | 176 | 353 | 0.92 | 6 |
| Into unused open plots | 2 | 9 | 16 | 29 | 116 | 181 | 353 | 0.75 | 7 |
| Composite Mean Score = 1.65 | | | | | | | | | |

*Key: A = Always; O = Often; S = Sometimes; R = Rarely; N = Never; U = Unspecified; MS = Mean Score; and R* = Ranking.*

**Table 5.**
*Waste storage material and frequency.*

were not specific. The respondents who indicated that they separate the waste, however, stated that they separate them into components of bottles, glasses, plastics, cans, boxes, cardboard and papers.

### *4.2.4 Waste disposal system and frequency*

**Table 7** shows that the disposal of waste into the Municipal waste truck with a mean score of 4.26 is often used as the means of waste disposal, while community bins (MS 1.82) are rarely used, and others such as recycling facilities, empty plots, landfill sites, abandoned houses and others are generally never used as their mean score is between 0.54 and 0.82. Besides, the composite mean of the waste disposal location stands at 1.58.

| Separate waste | Number of households | % |
|---|---|---|
| No | 323 | 91.5 |
| Unspecified | 11 | 3.1 |
| Yes | 19 | 5.4 |
| Total | 353 | 100.0 |

**Table 6.**
*Waste separation.*

| Waste disposal location | Response | | | | | | | | |
|---|---|---|---|---|---|---|---|---|---|
| | A | O | S | R | N | U | Total | MS | R* |
| Municipal waste truck | 261 | 41 | 9 | 1 | 5 | 36 | 353 | 4.26 | 1 |
| Community bins | 87 | 8 | 18 | 31 | 59 | 150 | 353 | 1.82 | 2 |
| Recycling facilities | 5 | 9 | 23 | 27 | 105 | 184 | 353 | 0.82 | 3 |
| Empty plots | 3 | 9 | 6 | 27 | 116 | 192 | 353 | 0.68 | 4 |
| Landfill sites | 1 | 10 | 8 | 24 | 119 | 191 | 353 | 0.67 | 5 |
| Abandoned houses | 5 | 5 | 4 | 12 | 136 | 191 | 353 | 0.62 | 6 |
| Others | 0 | 0 | 1 | 1 | 14 | 337 | 353 | 0.54 | 7 |
| Composite Mean Score = 1.58 | | | | | | | | | |
| *Key: A = Always; O = Often; S = Sometimes; R = Rarely; N = Never; U = Unspecified; MS = Mean Score; and R* = Ranking.* | | | | | | | | | |

**Table 7.**
*Domestic waste disposal system and frequency.*

When further asked about the frequency of the collection, as revealed in **Table 8**, the respondents indicated that the municipality is the main waste collector with 344 Nr (97.5%) indicating such while 5Nr (1.4%) indicated other and 4 Nr (1.1%) was unspecified.

### *4.2.5 Waste collection point and frequency of collection*

From the mean score ranking as presented in **Table 9**, door-to-door collection (MS 2.91) ranked the first in waste collection types, followed by community waste collection point (MS 1.95), while the collection of waste anywhere it is dumped (MS 0.70) and others (MS 0.20) ranks third and fourth respectively.

| Collector and place | Number of Households | % |
|---|---|---|
| Municipality | 344 | 97.5 |
| Other | 5 | 1.4 |
| Unspecified | 4 | 1.1 |
| Total | 353.0 | 100.0 |

**Table 8.**
*Waste collector.*

| Collection type | Frequency | | | | | | | | |
|---|---|---|---|---|---|---|---|---|---|
| | D | W | F | M | N | U | Total | MS | R* |
| Door-to-door collection | 5 | 241 | 7 | 1 | 15 | 84 | 353 | 2.91 | 1 |
| Community waste collection point | 2 | 155 | 8 | 6 | 22 | 160 | 353 | 1.95 | 2 |
| Collect waste anywhere it is dumped | 4 | 36 | 6 | 9 | 48 | 250 | 353 | 0.70 | 3 |
| Other | 0 | 12 | 3 | 2 | 9 | 327 | 353 | 0.20 | 4 |

*Key: D = Daily; W = Weekly; F = Fortnightly; M = Monthly; N = Never; U = Unspecified; MS = Mean Score; and R* = Ranking.*

**Table 9.**
*Waste collection point and frequency of collection.*

| Status | Number of households | % |
|---|---|---|
| No | 333 | 94.3 |
| Unspecified | 9 | 2.5 |
| Yes (government help) | 5 | 1.4 |
| Yes (no government help) | 6 | 1.7 |
| Total | 353 | 100.0 |

**Table 10.**
*Pay for waste removal.*

#### *4.2.6 Pay for waste removal*

As shown in **Table 10**, the number of respondents who indicated that they do not pay for waste removal is 333 (94.3%), and 9 (2.5%) respondents did not specify. Of the number that said that they do pay for waste removal, 5 Nr (1.4%) indicated that the government helps them, and 6 Nr (1.7%) stated that they receive no help from the government.

## 5. Discussion of findings

From the findings, it can be generally inferred that the respondents practice effective waste management in line with the provision of the local municipality and the municipality also fulfils its responsibility of waste collection.

To underscore the representativeness of the study, the findings show that both genders of females and males participated in the study, although the data stated that there were more women than men, and this is supported by Knoema [58], who indicated that there are more women than men in South Africa, with a ratio of 97 men to 100 women. Also, this finding is supported by data from UNDP [59] on South Africa, which states that 51.5 per cent are female, and 48.5 per cent are male. Furthermore, Arcgis [60] stated that the average South African household size in 2019 was 3.3 people, whereas the study revealed an average of 3.70 persons per household, a size within the same range, reiterating the validity of the findings. Literature has, however, indicated that the generation of waste which in turn dictates the waste management practice, is affected not only by the number of people but also by other factors like population structure or way of living and female to male ratio [61].

The survey identified that female household headship was in the majority, and literature indicates that female headship has been on the rise in South Africa, as recorded by the 10-yearly census data on female headship and income [62]. This assertion of [62] gives credence to the finding of this study. Furthermore, Posel [63] observed that the average age of South African heads of households was between 44 and 51 years and the average age of the participants from Ekuphumleni township was 46.32 years. While Anbazu [64] indicate that household heads influence the choice of waste management practice, Uma [65] further observed that many female-headed households utilise informal refuse disposal systems rather than male-headed households.

Using household members that are 20 years and above, the study observed that 12.5% of the respondents indicated that they have no schooling, but the finding of the waste practice showed that they have a good practice in consonance with Chikowore [66]. This finding is also supported by the observation of Mngomezulu [14] that the level of education has no association with waste management practices, but environmental education and a lack of information do.

The study observed that over 60% of the respondents are earning below R4000, putting them in the low-income strata of society. There is much scientific literature on the association of socioeconomic indices such as income with solid waste generation, but there are inconsistencies in the other literature findings. For example, Khan [67] stated that income significantly influences solid waste generation and management. Porpino [68] concluded that low-income households generate more waste, while Omolayo [69] concluded that higher-income households generate more waste than lower-income ones. Machate [70] observed that the causative factor is income, and that waste generation increases as income increases. Namlis [71] posited that the association was dependent on the development stage of a country and hypothesised that as income rises in emerging nations, so would solid waste generation; however, as income increases in advanced economies, waste generation significantly reduced. From the preceding, as expected, the waste generation in this community depended on other influencing factors asides from income.

Waste paper, cans, used plastics, and bottles were the major waste components generated in the township, and these wastes are recyclable. According to Chen [72], estimating the waste types generated and their management method can be useful for predicting future waste management trends. Nineteen case study of municipal solid waste in developing countries, as documented by Troschinetz [73], produces by average recyclable content of 55%. Such organic content includes food waste, paper and paper materials, human waste, bio-degradable plastic, and landscape and pruning waste, among others [9]. The data collected aligns with the literature on the typology of waste generated in developing countries and can be used to predict future trends and waste management systems.

Plastic bags and home garbage cans are indicated to be mostly used by the respondents as domestic storage materials, and according to the documentation of Yoada [74], the two most common storage items for domestic solid waste in Accra, Ghana, were plastic bins, baskets polythene bags, paper boxes and old buckets. Gumbi [75] also indicated that residents' major types of containers to dispose of waste collected by the municipality ranged from plastic bags to metal bins and plastic bins. The findings of this study show that plastic bag is the popular waste storage material in the township and aligns with other works of literature.

With over 90% indicating that they do not separate their waste, this finding aligns with the general apathy towards waste separation in developing countries where waste separation is uncommon, as observed by Ferronato [12]. According to Babaei [76], while initiatives to strengthen solid waste management in emerging economies have primarily focused on cost-effective practices such as separation, source reduction and recycling, their implementations have experienced social opposition because of low awareness and willingness to participate. Matete [77] also indicates that separation at source, among other things, is not yet accorded a top priority in line with regulatory and legal requirements in South Africa. Hence, it can be inferred that the respondent does not practice waste separation as part of their waste disposal practice.

The municipal waste truck always collects the waste from the respondents weekly from door-to-door as indicated by the respondents. According to the South African legislative provision, the municipality is responsible for solid waste management [78]. This agrees with literature from developed or emerging economies, such as Indonesia [79], Ghana [80], Colombia [81], Turkey [82], South Africa [52], USA [83] and the United Kingdom [84] among others where the municipal truck is the main collector of waste, but the difference is the frequency and efficiency. The finding of this study, where residents indicated that municipal truck comes to remove their wastes, confirms what the literature indicates.

However, Statistics South Africa [85] observed a lack of SWM services in South Africa, with only 66% of the population receiving waste collection services from municipalities or private companies through municipal contracts. With this background, Hlahla [86] indicated that South Africa has a variety of waste collection systems designed to accommodate the unique conditions of a peri-urban community, one of which is door-to-door collection by municipal truck, which is the collection practice in Ekuphumleni.

With over 94% of the respondents stating that they were not willing to pay for waste management services, the study aligns with Omolayo [87], who observed that socio-economic factors such as income level affect households' willingness to pay for waste management in South Africa. Therefore, the respondents' household income level could be inferred to be why the respondents are unwilling to pay for such services.

## 6. Conclusion and recommendations

The United Nations'2030 target continues to prioritise environmental sustainability. As a result, various levels of government in South Africa have implemented various waste disposal avenues for the populace, but there have been reported inefficiency of these program(s) in many parts of the nation. As a result, we investigated waste management practices in the South African township of Ekuphumleni using primary data.

The descriptive statistics findings show an average of 3.70 people (the ratio of females to males is 2.12 to 1.58) per household, and the average household head age of 46 years was recorded. In addition, about 87% of the household heads had formal education and over 60% with a monthly income estimated at below ZAR4000/USD200. The findings further identified that most of the waste generated by the township is recyclable, and that the main disposal method is storing the waste at the household level with some form of plastic material, which is collected weekly at their doors by the municipal truck. Many households perceive littering and the dumping of refuse anywhere as an environmental problem that requires drastic measures for its control or eradication.

Recycling has a mean score of 0.82, indicating the need for sensitisation programmes and incentives to increase household participation in recycling waste products.

Almost none of the respondents pay for waste, which can be attributed to some or all the socioeconomic factors, particularly household income.

The study's findings imply that the waste management practices of the residents of Ekuphumleni township do not fully align with the sustainable waste management practices of reducing, reusing, recycling and recovering, although a high volume of their waste is recyclable. The waste collected by the municipal trucks end up in landfills, thus contributing to greenhouse gas emission and pollution of the groundwater system.

Therefore, this study concludes that there is a need for sustainable waste management practices in the township. This is achievable by raising awareness and educating the residents on the need for sustainable waste management practices of reducing, reusing, recycling and recovering towards a circular economy. Towards this, the South African government must intensify its efforts on poverty alleviation interventions to improve the socioeconomic status of households and environmental sensitisation programmes through adequate citizen education to facilitate the achievement of zero waste.

## Acknowledgements

We acknowledge the efforts and hard work of the field survey teams who assisted with data collection under the very trying COVID-19 conditions. The cooperation and kind assistance of the Ndlambe Municipal Officials and the residents of Ekuphumleni and Dr. Kabundu for assistance with the statistical analysis. The Council for Scientific and Industrial Research (CSIR) and Department of Science and Technology (DST) in South Africa funded the research as part of the Ecosun Village Demonstration Project.

## Conflict of interest

The authors declare no conflict of interest.

## Author details

Ayo Adeniran[1*], Lorato Motsatsi[2], Sijekula Mbanga[1], Emma Ayesu-Koranteng[1] and Winston Shakantu[1]

1 Nelson Mandela University, Gqeberha, South Africa

2 Council for Scientific and Industrial Research (CSIR), Pretoria, South Africa

*Address all correspondence to: ayoadeniran111@gmail.com

## References


[1] Karri RR, Ravindran G, Dehghani MH. Wastewater—Sources, toxicity, and their consequences to human health. In: Soft Computing Techniques in Solid Waste and Wastewater Management. Amsterdam, Netherland: Elsevier; 2021. pp. 3-33

[2] Adeniran AA, Adewole AA, Olofa SA. Impact of solid waste management on ado Ekiti property values. Civil and Environmental Research. 2014;**6**(9):29-35

[3] Al-Taai SH. Solid waste: A study of its concept, management methods, and environmental impacts. IOP Conference Series: Earth and Environmental Science. 2022;**1002**(1):012007

[4] Ajibade FO, Adelodun B, Lasisi KH, Fadare OO, Ajibade TF, Nwogwu NA, et al. Environmental pollution and their socioeconomic impacts. In: Ajay K, Vipin KS, Pardeep S, Virendra KM, editors. Microbe-Mediated Remediation of Environmental Contaminants. 1st ed. Cambridge, UK: Woodhead Publishing; 2021. pp. 321-354

[5] Sharafi M, Atashkar S, Massahi T. Assessing the quantity of solid waste produced in Kermanshah School of Public Health: Challenges and suggestions. International Journal of Health and Life Sciences. 2021;7(2):1-6

[6] Hoornweg D, Bhada-Tata P. What a Waste: A Global Review of Solid Waste Management. Washington, USA: World Bank; 2012

[7] Demirbas A. Waste management, waste resource facilities and waste conversion processes. Energy Conversion and Management. 2011;**52**(2):1280-1287

[8] Chand S, Shastry CS, Hiremath S, Joel JJ, Bhat CK, Mateti UV. Water, sanitation, hygiene and biomedical waste disposal in the healthcare system: A review. Biomedicine. 2020;**40**(1):14-19

[9] Abdel-Shafy HI, Mansour MS. Solid waste issue: Sources, composition, disposal, recycling, and valorisation. Egyptian Journal of Petroleum. 2018;**27**(4):1275-1290

[10] U.N. Sustainable Development Goals [Internet]. Available from: http://www.un.org/sustainabledevelopment/sustainable-development-goals/ [Accessed: May 25, 2022]

[11] Wilson DC, Rodic L, Modak P, Soos R, Carpintero A, Velis K, et al. Global waste management outlook. UNEP. 2015

[12] Ferronato N, Torretta V. Waste mismanagement in developing countries: A review of global issues. International Journal of Environmental Research and Public Health. 2019;**16**(6):1060

[13] Ferronato N, Torretta V, Ragazzi M, Rada EC. Waste mismanagement in developing countries: A case study of environmental contamination. UPB. Scientific Bulletin. 2017;**79**(2):185-196

[14] Mngomezulu SK, Mbanga S, Adeniran AA, Soyez K. Factors influencing solid waste management practices in joe Slovo township, Nelson Mandela Bay. Journal of Public Administration. 2020;**55**(3):400-411

[15] Adeniran AA, Shakantu W. The health and environmental impact of plastic waste disposal in south African townships: A review. International Journal of Environmental Research and Public Health. 2022;**19**(2):779

[16] Guerrero LA, Maas G, Hogland W. Solid waste management challenges for cities in developing countries. Waste Management. 2013;**33**(1):220-232

[17] Giusti L. A review of waste management practices and their impact on human health. Waste Management. 2009;**29**(8):2227-2239

[18] Okwesili J, Iroko C. Urban solid waste management and environmental sustainability in Abakaliki urban. Nigeria. European Scientific Journal. 2016;**12**(23):155-183

[19] Campos-Alba CM, Garrido-Rodríguez JC, Plata-Díaz AM, Pérez-López G. The selective collection of municipal solid waste and other factors determining cost efficiency. An analysis of service provision by spanish municipalities. Waste Management. 2021;**134**:11-20

[20] Steenmans K, Malcolm R. Transitioning towards circular systems: Property rights in waste. Journal of Property, Planning and Environmental Law. 2020;**12**(3):219-234. DOI: 10.1108/JPPEL-03-2020-0018

[21] Thomas S. Waste, marginal property practices and the circular economy. Journal of Property, Planning and Environmental Law. 2020;**12**(3):203-218. DOI: 10.1108/JPPEL-02-2020-0012

[22] Hannon J, Zaman AU. Exploring the phenomenon of zero waste and future cities. Urban Science. 2018;**2**(3):90

[23] Wiprächtiger M, Rapp M, Hellweg S, Shinde R, Haupt M. Turning trash into treasure: An approach to the environmental assessment of waste prevention and its application to clothing and furniture in Switzerland. Journal of Industrial Ecology. 2022;**2022**(26):1-17. DOI: 10.1111/jiec.13275

[24] van Ewijk S, Stegemann JA. Recognising waste use potential to achieve a circular economy. Waste Management. 2020;**105**:1-7

[25] Awasthi AK, Li J, Koh L, Ogunseitan OA. Circular economy and electronic waste. Nature Electronics. 2019;**2**(3):86-89

[26] Inglezakis VJ, Moustakas K. Household hazardous waste management: A review. Journal of Environmental Management. 2015;**150**:310-321

[27] Raubenheimer K, McIlgorm A. Can the Basel and Stockholm conventions provide a global framework to reduce the impact of marine plastic litter? Marine Policy. 2018;**96**:285-290

[28] Oladunjoye MA, Adeniran MA. Towards efficient Management of Solid Wastes: An overview of dumpsites in Ibadan Metropolis, southwestern Nigeria. The Journal of Solid Waste Technology and Management. 2021;**47**(2):271-284

[29] Flores CAR, da Cunha AC, Cunha HFA. Solid waste generation indicators, per capita, in Amazonian countries. Environmental Science and Pollution Research. 2022;**29**(22):33138-33151

[30] Sun L, Chen J, Li Q, Huang D. Dramatic uneven urbanisation of large cities throughout the world in recent decades. Nature Communications. 2020;**11**(1):1-9

[31] Amaral C, Pedro MI, Ferreira DC, Marques RC. Performance and its determinants in the Portuguese municipal solid waste utilities. Waste Management. 2022;**139**:70-84

[32] Oteng-Ababio M. E-waste: An emerging challenge to solid waste management in Ghana. International

Development Planning Review. 2010;**32**(2):191-206

[33] Khan S, Anjum R, Raza ST, Bazai NA, Ihtisham M. Technologies for municipal solid waste management: Current status, challenges, and future perspectives. Chemosphere. 2022;**288**:132403

[34] Xu L, Ling M, Wu Y. Economic incentive and social influence to overcome household waste separation dilemma: A field intervention study. Waste Management. 2018;**77**:522-531

[35] Jaunich MK, Levis JW, DeCarolis JF, Gaston EV, Barlaz MA, Bartelt-Hunt SL, et al. Characterisation of municipal solid waste collection operations. Resources, Conservation and Recycling. 2016;**114**:92-102

[36] Idowu IA, Atherton W, Hashim K, Kot P, Alkhaddar R, Alo BI, et al. An analyses of the status of landfill classification systems in developing countries: Sub Saharan Africa landfill experiences. Waste Management. 2019;**87**:761-771

[37] Association for Water and Rural Development [AWARD]. South Africa Is Drowning in Its Own Waste-Are Our Regulators Taking This Crisis Seriously? [Internet]. Available from: http://award.org.za/index.php/2019/02/01/south-africa-is-drowning-inits-own-waste-are-our-regulators-taking-this-crisis-seriously/#:~{}:text=South%20Africans%20generate%20roughly%2054,90%25%20is%20landfilled%20or%20dumped [Accessed: May 5, 2022]

[38] Muhammad LJ, Badi I, Haruna AA, Mohammed IA. Selecting the best municipal solid waste management techniques in Nigeria using multi-criteria decision-making techniques. Reports in Mechanical Engineering. 2021;**2**(1):180-189

[39] Maria C, Góis J, Leitão A. Challenges and perspectives of greenhouse gases emissions from municipal solid waste management in Angola. Energy Reports. 2020;**6**:364-369

[40] Adusei-Gyamfi J, Boateng KS, Sulemana A, Hogarh JN. Post COVID-19 recovery: Challenges and opportunities for solid waste management in Africa. Environmental. Challenges. 2022;**6**(2022):1-9

[41] Röhlig KJ. Waste management policy and strategy. Nuclear Waste. 2022;**4**(2022):1-12

[42] Adeniran AA, Shakantu W. The role of virtual reality in planning for municipal solid waste management. Sustainability Handbook. 2021:130

[43] Godfrey L, Ahmed MT, Gebremedhin KG, Katima JH, Oelofse S, Osibanjo O, et al. Solid Waste Management in Africa: Governance Failure or Development Opportunity? Regional Development in Africa2019. p. 235

[44] United Nations Environment Programme [UNEP]. Enforcement of Environmental Law: Good Practices from Africa, Central Asia, ASEAN Countries and China [Internet]. Available from: http://wedocs.unep.org/bitstream/handle/20.500.11822/9968/enforcement-environmentallaws.pdf?sequence=1&isAllowed=y [Accessed: March 21, 2022]

[45] David VE, John Y, Hussain S. Rethinking sustainability: A review of Liberia's municipal solid waste management systems, status, and challenges. Journal of Material Cycles and Waste Management. 2020;**22**(5):1299-1317

[46] Azage M. Healthcare waste management practices among healthcare workers in healthcare facilities of Gondar

town. Northwest Ethiopia. Health Science Journal. 2013;7(3):0-0

[47] Ndum AE. Bottom-up Approach to Sustainable Solid Waste Management in African Countries (Doctoral Dissertation). Universitätsbibliothek der BTU Cottbus; 2013

[48] Mir IS, Cheema PPS, Singh SP. Implementation analysis of solid waste management in Ludhiana city of Punjab. Environmental Challenges. 2021;**2**:100023

[49] Stern N, Valero A. Innovation, growth and the transition to net-zero emissions. Research Policy. 2021;**50**(9):104293

[50] Farooq A, Haputta P, Silalertruksa T, Gheewala SH. A framework for the selection of suitable waste to energy technologies for a sustainable municipal solid waste management system. Frontiers in Sustainability. 2021;**2**:681690

[51] Wilson DC, Rodic L, Scheinberg A, Velis CA, Alabaster G. Comparative analysis of solid waste management in 20 cities. Waste Management & Research. 2012;**30**(3):237-254

[52] Simatele DM, Dlamini S, Kubanza NS. From informality to formality: Perspectives on the challenges of integrating solid waste management into the urban development and planning policy in Johannesburg, South Africa. Habitat International. 2017;**63**:122-130

[53] Muheirwe F, Kombe W, Kihila JM. The paradox of solid waste management: A regulatory discourse from sub-Saharan Africa. Habitat International. 2022;**119**:102491

[54] South African Waste Information Centre [SAWIC]. Waste Policy and Legislature, 2015 [Internet]. Available from: http://www.sawic.org.za/?menu=13 [Accessed: May 6, 2022]

[55] Pimentel JL, Pimentel JL. Some biases in Likert scaling usage and its correction. International Journal of Science: Basic and Applied Research (IJSBAR). 2019;**45**(1):183-191

[56] Sarrafzadeh M, Martin B, Hazeri A. Knowledge management and its potential applicability for libraries. Library Management. 2010;**31**(3):198-212. DOI: 10.1108/01435121011027363

[57] Fink A. How to Conduct Surveys: A Step-by-Step Guide. Los Angeles, USA: Sage Publications; 2015

[58] Knoema. South Africa Male to female ratio, 1950-2018. [Internet]. Available from: https://knoema.com/atlas/South-Africa/topics/Demographics/Population/Male-to-female-ratio [Accessed: February 9, 2022]

[59] United Nations Development Programme [UNDP]. Human Development Indices and Indicators 2018 Statistical Update [Internet]. Available from: https://hdr.undp.org/sites/default/files/2018_human_development_statistical_update.pdf [Accessed: February 9, 2022]

[60] Arcgis. Average Household Size in South Africa [Internet] 2022. Available from: https://www.arcgis.com/home/item.html?id=582208ececa2424ab6e387d9cdcf01e3#:~:text=Description- [Accessed: February 9, 2022]

[61] Talalaj IA, Walery M. The effect of gender and age structure on municipal waste generation in Poland. Waste Management. 2015;**40**:3-8

[62] Sambo, N. and Kennedy, P. Womxn in Big Data South Africa: Female-Headed Households in South Africa

[Internet]. Available from: https://www.womeninbigdata.org/womxn-in-big-data-south-africa-female-headed-households-in-south-africa/ [Accessed: July 22, 2022]

[63] Posel DR. Who are the heads of household, what do they do, and is the concept of headship useful? An analysis of headship in South Africa. Development Southern Africa. 2001;**18**(5):651-670

[64] Anbazu JK, Ocloo KA, Oduro-Ofori E. Determinants of the choice of disposal methods among residents in urban areas in southern Ghana. Journal of the Air & Waste Management Association. 2022:just-accepted

[65] Uma KE, Nwaka ID, Nwogu MU, Obidike PC. What are the triggers of household decision-making on waste disposal choices? A gender-differentiated analysis. Heliyon. 2020;**6**(12):e05588

[66] Chikowore N. Factors influencing household waste management practices in Zimbabwe. Journal of Material Cycles and Waste Management. 2021;**23**(1):386-393

[67] Khan D, Kumar A, Samadder SR. Impact of socioeconomic status on municipal solid waste generation rate. Waste Management. 2016;**49**:15-25

[68] Porpino G, Parente J, Wansink B. Food waste paradox: Antecedents of food disposal in low-income households. International Journal of Consumer Studies. 2015;**39**(6):619-629

[69] Omolayo Y, Feingold BJ, Neff RA, Romeiko XX. Life cycle assessment of food loss and waste in the food supply chain. Resources, Conservation and Recycling. 2021;**164**:105119

[70] Machate M. Reflections on the influence of family demographics on food waste generation among the city of Tshwane households, Republic of South Africa. In: Saleh H, editor. Strategies of Sustainable Solid Waste Management. London: IntechOpen; 2021

[71] Namlis KG, Komilis D. Influence of four socioeconomic indices and the impact of economic crisis on solid waste generation in Europe. Waste Management. 2019;**89**:190-200

[72] Chen DMC, Bodirsky BL, Krueger T, Mishra A, Popp A. The world's growing municipal solid waste: Trends and impacts. Environmental Research Letters. 2020;**15**(7):074021

[73] Troschinetz AM, Mihelcic JR. Sustainable recycling of municipal solid waste in developing countries. Waste Management. 2009;**29**(2):915-923

[74] Yoada RM, Chirawurah D, Adongo PB. 2014. Domestic waste disposal practice and perceptions of private sector waste management in urban Accra. BMC Public Health. 2014;**14**(1):1-10

[75] Gumbi SE. Current Waste Management and Minimisation Patterns and Practices: An Exploratory Study on the Ekurhuleni Metropolitan Municipality in South Africa (Doctoral Dissertation). Johannesburg: University of South Africa; 2015

[76] Babaei AA, Alavi N, Goudarzi G, Teymouri P, Ahmadi K, Rafiee M. Household recycling knowledge, attitudes and practices towards solid waste management. Resources, Conservation and Recycling. 2015;**102**:94-100

[77] Matete N, Trois C. Towards zero waste in emerging countries–A south African experience. Waste Management. 2008;**28**(8):1480-1492

[78] De Villiers NF, Van As HJ, Botha JC. The enforceability of the by-laws of district municipalities on local municipalities: The case of solid waste disposal. Stellenbosch Law Review. 2020;**31**(2):315-343

[79] Chaerul M, Mulananda AM. Minimisation of municipal solid waste transportation route in West Jakarta using Tabu search method. IOP Conference Series: Earth and Environmental Science. 2018;**148**(1):012026

[80] Sulemana A, Donkor EA, Oduro-Kwarteng S. Efficiency of municipal solid waste collection systems in Ghana. The Journal of Solid Waste Technology and Management. 2020;**46**(1):58-65

[81] Ortiz-Rodriguez OO, Rivera-Alarcon HU, Villamizar-Gallardo RA. Evaluation of municipal solid waste by means of life cycle assessment: Case study in the South-Western region of the Department of Norte De Santander. Colombia. Environmental Engineering & Management Journal (EEMJ). 2018;**17**(3):611-619

[82] Korkut NE, Yaman C, Küçükağa Y, Jaunich MK, Demir İ. Greenhouse gas contribution of municipal solid waste collection: A case study in the city of Istanbul, Turkey. Waste Management & Research. 2018;**36**(2):131-139

[83] Mahmood I, Zubairi JA. Efficient waste transportation and recycling: Enabling technologies for smart cities using the internet of things. IEEE Electrification Magazine. 2019;**7**(3): 33-43

[84] Bala A, Raugei M, Teixeira CA, Fernández A, Pan-Montojo F, Fullana-i-Palmer P. Assessing the environmental performance of municipal solid waste collection: A new predictive LCA model. Sustainability. 2021;**13**(11):5810

[85] Statistics South Africa. Census 2016: Community Profiles [Internet]. 2016. Available from: http://interactive.statssa.gov.za/superweb/login.do [Accessed: July 22, 2022]

[86] Hlahla S, Goebel A, Hill TR. Green economy: A strategy to alleviate urban poverty and safeguard the environment? KwaZulu-Natal, South Africa. Urban Forum. 2016;**27**(1):113-127. DOI: 10.1007/s12132-015-9263-7

[87] Omotayo AO, Abeeb BO, Adebola SD, Adebayo IO, Kehinde OA. What drives households' payment for waste disposal and recycling behaviours? Empirical evidence from South Africa's general household survey. International Journal of Environmental Research and Public Health. 2020;**17**(19):7188

Chapter 5

# Health Impacts of Poor Solid Waste Management in the 21st Century

*Parin Somani*


## Abstract

Industries and households globally accumulate solid waste at a heightened rate, due to increased global population, food and essentials to survive. Although steps have been taken in recent years to manage waste efficiently, through utilizing municipal waste collection centres and taken to landfills and dumps; solid waste is presenting impacts on human health. This is particularly prevalent within developing countries. This study aims to understand the health impacts of poor waste management in the 21st century. A systematic review of white and gray literature sources is carried out. Results have revealed that poor solid waste management comprising of waste generated by human beings and animal activity, result in the following: a spread of infections and diseases through attracting rodents and other creatures, pollution from chemicals released through landfills and greenhouse gasses, plastic waste and respiratory diseases. Global societies should be educated on implementing appropriate strategies towards good solid waste management.


**Keywords:** solid waste, health, management, infection, diseases

## 1. Introduction

Industries and households globally accumulate solid waste at a heightened rate, due to increased global population, food and essentials to survive. Attempts have been made within governmental policies to attain sustainable societal development. It has been identified that through ensuring the implementation of sustainable solid waste management strategies, the United Nation's Sustainable Development Goals (SDG's) can be achieved. This includes SDG6 which is "Ensure access to water and sanitation for all" SDG 11 ensuring "Sustainable cities and communities" [1], SDG 13 "Take urgent action to combat climate change" [2], SDG 15 "Life on land protect, restore and promote sustainable use of terrestrial ecosystems, sustainably manage forests, combat desertification, and halt and reverse land degradation, and halt biodiversity loss" [3] and SDG 12 "Ensuring sustainable consumption and production patterns, which is key to sustain the livelihoods of current and future generations" [4]. Solid waste management facilitates a decrease in the number of finite resources that are used, the number or reusable materials in addition to reduced recycling to irradicate waste, decrease the amount of pollution, helping to save costs and improve the green economy. When this occurs, not only will there be a growth in economy, but individuals will also enjoy a better and healthier lifestyle.

It is predicted that by 2025 the global population will increase to eight billion and by 2050 there will be a further increase to 9.3 billion out of which approximately seventy percent of the population will reside in urban locations [5]. In addition, it is predicted that each individual will increase the amount of municipal solid waste they produce, due to urbanization and increased industrialization [6]. Numerous developing countries have remained underdeveloped pertaining to their solid waste management systems. Approximately ninety percent of residual waste within urban areas is dumped opposed to it being landfilled in the appropriate manner. Within India in the millennium, the MoEF declared solid waste management and handling guidelines through which waste management could be completed appropriately [7]. The rules created a pathway through which governmental authorities were able to devise and implement a feasible infrastructure through which solid waste could be collected, stored, segregated, transported, processed, and disposed [8]. Research has identified that after spending between twenty to fifty percent of their budget, only fifty to eighty percent of general waste is collected using eighty to ninety five percent on transportation and waste collection within developing countries [9]. **Figure 1** illustrates the most and least preferred steps of solid waste management.

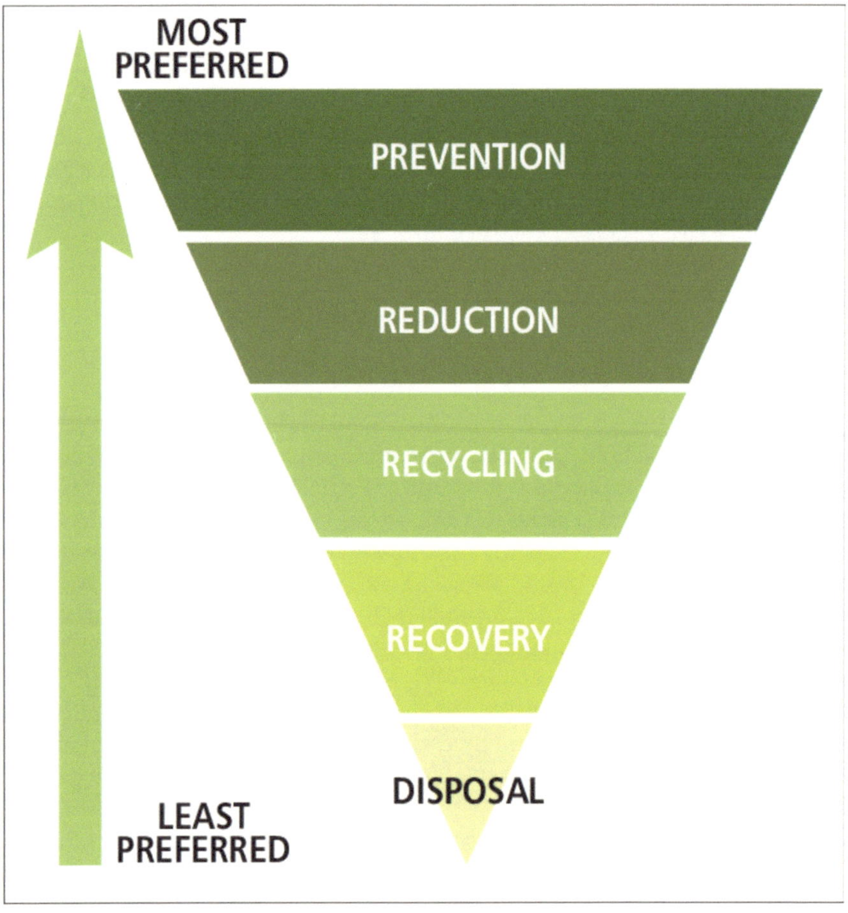


**Figure 1.**
*Most and least preferred steps of solid waste management [10].*

*DOI: http://dx.doi.org/10.5772/intechopen.1002812*

The most preferred step in the solid waste management hierarchy is prevention. This is because the least number of materials and resources are utilized to manufacture and design, ensuring minimal generation of waste towards a clean and friendly environment. The next step is reduction and refers to the refurbishment, cleaning, checking and repairing spare or whole items that are found. This step does not allow inappropriate waste products to enter the disposal system without being checked. Waste is collected during the interim process production, then returned to the source to facilitate the complete production processes, resulting in a reduction in waste generation. Recycling is the next step, during which materials can be extracted and reused into a new product or substance. Alternatively, organic waste can be composited, and soil fertility can be improved. The recovery process is not as desired, because it results in the anaerobic digestion and incineration with recovering energy. It also includes other useful materials like gasification and pyrolysis used to produce energy like power, heat and fuel. Thus, although it is not the most preferred step it ensures waste is converted into very useful sources of energy. The final step is disposal, and it is placed within the solid waste management hierarchy as the least preferred step. This is because it involves incineration without recovering any energy and also includes landfilling.

Although steps have been taken in recent years to manage waste efficiently, through utilizing municipal waste collection centers and taken to landfills and dumps; solid waste is presenting impacts on human health. This is particularly prevalent within developing countries. This study aims to understand the health impacts of poor waste management in the 21st century.

### 1.1 Objectives

This study aims to understand the health impacts of poor waste management in the 21st century.

## 2. Methodology

A systematic review through a well-planned literature search is implemented using manual and electronic databases. Literature sources are extracted from, analyzed, evaluated, and interpreted.

## 3. Results and discussion

Results have revealed that poor solid waste management comprising of waste generated by human beings and animal activity, result in the following: a spread of infections and diseases through attracting rodents and other creatures, pollution from chemicals released through landfills and greenhouse gasses, plastic waste and respiratory diseases.

### 3.1 A spread of infections and diseases

With more than two billion tons of municipal solid waste generated per year, it is vital to ensure that it is disposed of properly, minimizing harmful outcomes on public health. This is due to high contamination levels within the air, water and soil consequently affecting the lives of both adults and children. Waste deemed as hazardous, or the implementation of waste treatment that is unsafe like open burning, can create a harmful impact of employees and other individuals involved within the burning processes. In addition, local communities are subject to increased health risks with children and vulnerable adults are more at risk. When inadequate collection of waste is carried out, impacts upon the environment and marine pollution including blockage of water drains have been highlighted. Subsequently, floods and can occur and initiate vector-borne diseases like malaria, dengue and promote cholera [11]. According to the World Health Organization in a survey carried out in 2019, approximately fifty-four million tons of e-waste was produced per annum. They included televisions, phones, and computers. There is an expectation that by 2030 this will rise to seventy-five million tons.

A process of fermentation can occur due to improper disposal of organic waste, which promotes the ideal environment for microbial pathogens to "survive and thrive" [12]. This led to serious human health hazards because, when humans have direct contact with such solid waste, they can contract chronic ailments and infections. When solid waste is left unattended on roadsides and waste accumulates, the effects can be most harmful as they gradually become a breeding ground for the infestation of rats, mosquitoes and cockroaches. It is well known that rodents can contribute to food poisoning, Dengue and Malaria. Hence it is vital that solid waste is properly disposed of, to limit the number of pests carrying diseases and a reduction of the impact on the health of the larger public population.

When solid waste is not properly disposed of using scientific methods, the individuals most at risk are municipal workers and rag pickers. This includes when individuals have been exposed to pollutants and toxic substances, as this can result in potential blood infections and skin irritation. **Figure 2** provides a strategy to facilitate

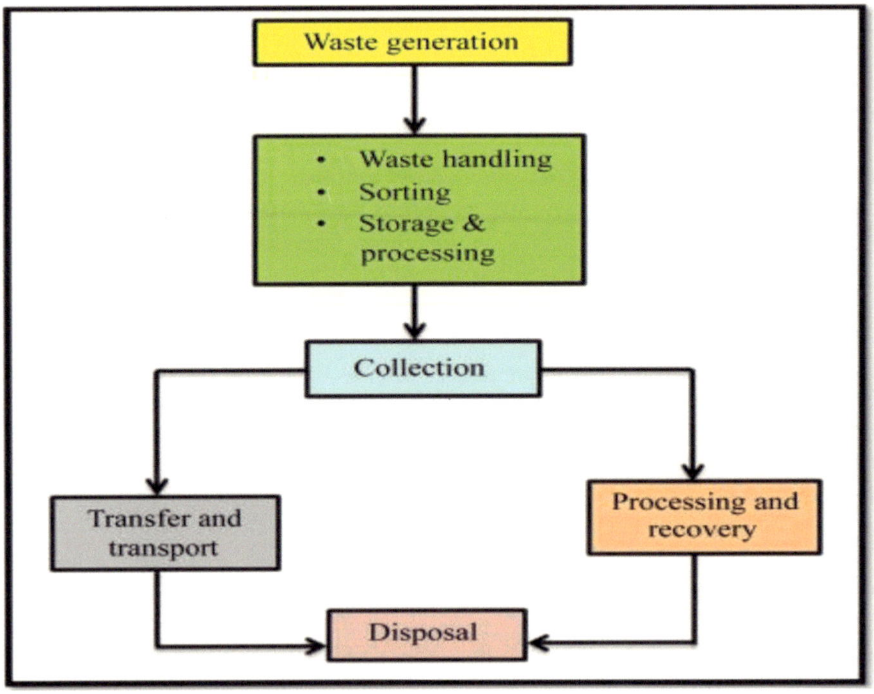


**Figure 2.**
*Sustainable strategy of organic waste management facilitation [13].*

sustainable management of organic waste. It includes the generation of waste, which should then be handled, sorted, stored, and processed appropriately. The waste should then be collected, processed, and recovered or transferred and transported, after which it should be correctly disposed of.

### 3.2 Pollution from chemicals

The local area economy has an impact upon the amount of waste composition, according to research. The use of packaged goods are likely to be utilized by high earners, for example plastic, glass, metals and paper. Hence, waste management practices are impacted and reflected through waste composition discrepancies [14]. Some waste is deemed as hazardous including medicines, pesticides and batteries which are mixed in with the municipal waste. In addition, fruit and vegetables are organic waste, while biomedical waste should not be mixed with municipal solid waste like blood-stained clothes, sanitary products, and disposable syringes, due to the risk of contamination and increased infection rates [15]. However, the most amount of organic waste is accumulated through domestic use, in contrast demolition produce and road sweeping constitutes to inert waste. Nevertheless, the composition of municipal solid waste can differ between cities [16].

Junkyards have vastly contributed to detrimental impacts on public health and the environment [17]. Anaerobic conditions are generated within open dumps leading to the production of methane which is a decomposition of biodegradable waste. However, the methane gas is a major contributing element in global warming and facilitating explosions and fires [18]. Other challenges contributing to poor health implications is the production of bad odor and generation of leaches that accumulate within the water reservoirs [19]. This is most notable within developing countries like India, when encountering hot weather conditions reaching 45°C [20]. In addition, respiratory diseases have been heightened due to waste being burnt uncontrollably, without the implementation of adequate controls. This has resulted in fine particles forming, consequently leading to smog and respiratory diseases [17]. Poor waste management has resulted in tremendous effects on public health including an increase in infections. This can involve bacterial infections, throat and nose inflammation, asthma, allergies, difficulties in breathing, and a reduced immunity [21].

Even within developed counties like the United Kingdom, research has revealed that large amounts of waste are being dumped illegally. This posses as a threat to human health and impacts environmental factors. This is because when chemicals that are hazardous, they can contaminate the air, soil and water, affecting human health, and endangering marine life and wildlife. Within a study carried out by The British Medical Bulletin, including individuals residing near a dumpsite, results revealed that the improper management of waste affected residents that lived close and far from the dumpsite. They reported symptoms like cholera, chest pains and diarrhea [22]. Although attempts are being made to achieve countries that are cleaner and greener, there is a need to ensure better solid waste management processes are in place. There are still individuals and companies that utilize unorthodox waste disposal companies and fly-tipping which is deemed to have numerous environmental and public consequences [23]. It has been revealed through a local government association analysis in the United Kingdom, that there has been a thirteen percent increase in cost in comparison to the previous year. This has affected the tax payer and land owner in order for fly-tipping to take place. In addition, due to a lack of awareness of the detrimental effects of waste materials, individuals continue to dispose of hazardous waste in fly-tipping. It was revealed that approximately more that forty tons of electric waste was produced every year and dumped illegally. Due to the mixture of chemicals, exposure to these resulted in harmful impacts to both humans and wildlife.

### 3.3 Plastic waste and respiratory diseases

It has been revealed that only approximately ten percent of waste generated is collected and disposed of adequately within suburban areas. Thereby contributing to environmental and public health risks. This has been linked to acute respiratory infections, more prevalent in children who reside near garbage dumps. Other symptoms reported included increased diarrhea within individuals [24].

Solid waste management requires a well-organized method of resource abstraction so that valuable resources can be found from the waste through a systematic process. In addition, low-income individuals can benefit from the position. Through this method, new material, energy, and nutrients can be found and recycled [25]. To achieve this, investment in solid waste management is required, which will facilitate novel research and development initiatives that can be devised and implemented, in addition to aid finding materials that can be recycled [14]. Within many developing countries like India there is a lack of organization and methodological process that can be followed to segregate waste in a community or within domestic use. It is left to the producers of the waste, or unorganized sectors to dispose of and manage the solid waste. Their lack of processes results in inefficient segregation and sorting, leading to insecurity and low-quality resource abstraction processes. This is because only valuable waste products are extracted due to higher financial gain [26].

## 4. Conclusion

Global societies should be educated on implementing appropriate strategies towards good solid waste management. The least preferred method for solid waste management in accordance with the 2016 MoUHA is landfills, due to the production of excess air pollution. Residents within communities living near to landfills may inhale toxic pollutants like methane and hydrogen sulphide that can result in major health challenges. In addition, the use of incinerators within solid waste management, have revealed "high toxic emission levels that can put human health at risk" [12]. In contrast the best method to ensure solid waste management is completed effectively is composting which individuals, and industries can practice in their own vicinities. This is deemed to reduce the burden of governing bodies, their communities, and local governments. Individuals should adhere to guidelines of segregating waste in the proper manner. This will allow composting to occur with the organic waste that is separated, treated then utilized as a fertilizer. This can be created from food that has been left over, waste accumulated from fruit, vegetables, and paper, all of which are used for domestic purposes. A variety of treatment options can be used to decompose organic waste. One of which can be found in a spray form that enables bacteria to decompose waste at a rapid rate resulting in a compost that is rich. It is up to everyone in society to protect each other and the planet from the harmful negative impacts of solid waste management.

It is important to remember that solid waste that is not disposed of in the correct manner can heighten mortality rates, cancer and can exacerbate reproductive health challenges. Thus, the management of solid waste is vital within both rural and urban areas. There is a need for policymakers, head of organizations and the public to devise innovative and environmentally friendly methods through which solid waste can be managed appropriately and efficiently. Depending on where the waste is generated, it can be segregated into categories like municipal solid waste, e-waste, or health waste. In effect any discarded materials, or rubbish can be referred to as solid waste. Approximately seventeen percent of e-waste was properly documented in 2019. This vast rise in e-waste and improper disposal of e-waste can result in several negative health implications relating to development and health particularly in young children. Hence, the importance of disposing all solid waste in an appropriate and safe manner and segregating them into different categories for easier management has become imperative.

It is essential to create awareness within global societies on challenges pertaining to solid waste management and the consequential detrimental health implications. Through this, individuals will have a better understanding of lasting damaging effects of solid waste management and solutions towards creating a sustainable progressive and safer future in the twenty first century and beyond.

## Conflict of interest

The authors declare no conflict of interest.

## Author details

Parin Somani
London Organisation of Skills Development, UK

*Address all correspondence to: drparinsomani@gmail.com

## References


[1] UN. UN Goal 11: Make Cities Inclusive, Safe, Resilient and Sustainable. Geneva: United Nations; 2023

[2] UN. UN 13 Take Urgent Action to Combat Climate Change and its Impacts. Geneva: United Nations; 2023

[3] Globalgoals, 15 Life on Land, The Global Goals. Available from: https://www.globalgoals.org/goals/15-life-on-land/ 2023.

[4] UN. UN Goal 12: Ensure Sustainable Consumption and Production Patterns. Geneva: United Nations; 2023

[5] OECD. Environmental Indicators. Development, Environmental Indicators. Development, Measurement and Use. Paris: Organisation for Economic Co-operation and Development; 2003

[6] PPCB. Status Report on Municipal Solid Waste in Punjab. Patiala: Punjab Pollution Control Board; 2010

[7] MoEF. Municipal Solid Waste (Management and Handling) Rules. New Delhi: Ministry of Environment and Forests; 2015

[8] Rana R, Ganguly R, Gupta AK. An assessment of solid waste management system in Chandigarh city India. Electronic Journal of Geotechnical Engineering. 2015;**20**:1547-1572

[9] Guerrero L, Maas G, Hogland W. Solid waste management challenges for cities in developing countries. Waste Management. 2013;**33**:220-232

[10] Sharma I. Solid Waste Management Hierarchy. 2019. Available from: https://ishaneesharma.home.blog/2019/05/09/solid-waste-management-hierarchy/ [Accessed: June 15, 2023].

[11] WHO. Compendium of WHO and Other UN Guidance on Health and Environment. Geneva: World Health Organization; 2023

[12] Kenekar A. Negative Effects of Improper Solid Waste Disposal on Human Health. 2021. [Online]. Available from: https://organicabiotech.com/negative-effects-of-improper-solid-waste-disposal-on-human-health/#:~:text=Water%20and%20Air%20Pollution&text=During%20the%20rainy%20season%2C%20-the,like%20cholera%2C%20diarrhoea%20and%20dysentery.

[13] Renu MP. Strategies toward sustainable management of organic waste. In: Techno-Economics and Life Cycle Assessment of Bioreactors. Elsevier; 2022. pp. 131-144

[14] Sridevi P, Modi M, Lakshmi M, Kesavarao L. A review on integrated solid waste management. International Journal of Engineering Scienes and Advanced Technology. 2012;**2**:1491-1499

[15] Bhalla B, Saini M, Jha MK. Effect of age and seasonal variations on leachate characteristics of municipal solid waste. International Journal of Research in Engineering and Technology. 2013;**2**:223-232

[16] Rawat M, Ramanathan AL, Kuriakose T. Characterization of municipal solid waste compost from selected Indian cities: A case study for its sustainable utilization. Environmental Protection. 2013;**4**:163-171

[17] Somani P, Navaneethan RD, Thangaiyan S. Integrated solid waste management in urban India: A mini review. Journal of Physics: Conference Series. 2021;**1913**:012084

[18] Somani P. Impact of climate change and covid-19 on health. Journal of Hunan University(Natural Sciences). 2021;**48**:184-193

[19] Srivastava R, Krishna V, Sonkar I. Characterization and management of municipal solid waste: A case study of Varanasi city, India. The International Journal of Current Research and Academic Review. 2014;**2**:10-16

[20] Dasgupta B, Yadav VL, Mondal MK. Seasonal characterization and present status of municipal solid management in Varanasi, India. Advances in Environmental Research. 2013;**2**:51-60

[21] CPCB. Management of Municipal Solid Waste Delhi. Delhi: Central pollution Control Board; 2000

[22] Rushton L. Health hazards and waste management. British Medical Bulletin. 2003;**68**:183-197

[23] Wade T. Effects of improper waste disposal on human health [Online]. 2018. Available from: https://www.speedyclearances.com/post/effects-of-improper-waste-disposal-on-human-health.

[24] UN-Habitat. Solid waste Management in the World's cities: Water and sanitation in the World's cities 2010. In: United Nations Human Settlements Programme. London, UK: Earthscan; 2010

[25] ISWA. Globalization and Waste Management Final Report from the ISWA Task Force. International Solid Waste Association; 2012

[26] Kaushal R, Varghese GK, Chabukdhara M. Municipal solid waste management in India-current state and future challenges: A review. International Journal of Engineering Science and Technology. 2012;**4**(4):1473-1489

Chapter 6

# Simple Is Better When Appropriate: An Innovative Approach to Biowaste Treatment Using Wild Black Soldier Fly

*Atinuke Chineme, Marwa Shumo, Getachew Assefa, Irene Herremans and Barry Wylant*

## Abstract

The acknowledgement that "technology is not a panacea" creates opportunities to dialog about appropriate technologies that keep the local context at the forefront of deriving solutions. The Black Soldier Fly (BSF) biowaste treatment method offers one such opportunity, and its simplistic adaptation is critical in locations with waste collection and management challenges. In this chapter, the importance of local context on viable waste solutions will be discussed with the applicability of appropriate technology strategies. First, the Black Soldier Fly waste treatment method will be distinguished as an appropriate technology for low-income communities. Then, a brief history of the nascent BSF method will be traced, followed by the production scales favored by world regions. Finally, an open BSF bioreactor case study will be introduced and analyzed.

**Keywords:** appropriate technology, Black Soldier Fly, biowaste, community-based, open system, wild Black Soldier Fly, bioreactor, waste treatment, co-production, co-design

## 1. Introduction

The expert consensus is that municipal solid waste management (MSWM) is a global and complex task ([1], p. 9); ([2], p. 10); [3] that requires modified management techniques which depend on several factors. MWSM aims to reduce waste volume through reuse, recycling, and recovery while mitigating pollution risks. Therefore, waste management considerations include but are not limited to population, topography, climate, culture, and purchasing power (budget) of the community, city, or region. MSWM is an expensive venture and is the single highest budget line item for some low-income municipalities [4]. Governments face the challenging issue of recovering expenses from low-income communities, which is further compounded by limited municipal budgets. These obstacles make it arduous to eradicate hazardous unregulated waste disposal practices like open burning and open dumping, which are easily accessible but pose significant risks [5]. Low-income locations face additional complexities and hurdles, which include the implementation and consequent breakdown of inappropriate technologies [6], user-rejected top-down MSWM designs [7], vehicular inaccessibility due to cursory urban planning, and the scarcity of government-run facilities [8]. All these factors contribute to MSWM remaining an unresolved challenge [9], particularly in low-income regions. As such, there is a consensus that a solution to the solid waste management challenge in low-income countries must be locally developed to remain long-lasting [4, 10–13]. Managing MSW involves more than just a technological system that aids in the handling and disposal of waste [14]. It incorporates the socio-economic, cultural, and operating conditions of its environment. This is where appropriate technologies come in. Appropriate technology (AT) concentrates on co-designing solutions focused on the "basic needs of water, sanitation, and agriculture" with the communities that are to be served [15]. The following section will explore the use of AT in infrastructure projects and its application in MSWM.

### 1.1 Appropriate technology as an alternative in solid waste management

Locally developed technologies that empower communities and build the capacities of local grassroots are conventionally deemed appropriate technologies [15]. These technologies provide necessities like water, clean air, safe and healthy food, and sanitation. "Some of the tenets generally applicable to ATs include: require little capital, utilize local materials and resources, be relatively labour intensive, be small scale and be affordable" [15]. Organizations such as the Farallon Institute in the United States stress that assistive technology (AT) should not only be affordable but also easily accessible and simple to maintain, with the ability to adapt to small-scale applications; the "socio-cultural and geographical contexts" of the location must constrain the technology. While modern and ultra-modern technologies have their place, they are not always the solution to every problem. As illustrated by Endresen and Hesselberg [16], the current modern technology deployed in mining diamonds in Botswana is inappropriate because it restricts employment instead of satisfying the urgent societal need for employment. On the other hand, using labour-intensive methods to mine diamonds (traditional methods) is equally inappropriate [16].

Appropriate technology can be traced to the post-World War II era when criticism grew over the inefficiency of technical aid support to redeveloping and developing countries [17].

> *To a significant degree, the American aid programs, and those of other developed nations, were captive to the notion that ideally all countries should follow the same pattern of industrialization, in both urban and rural settings.... as it turned out, such efforts ignored or misunderstood local environments, both natural and cultural.... Dams that destroyed fisheries, dual economies that privileged local elites and machinery that lay idle because of a lack of fuel or maintenance eventually led to the realization that many technologies that might be useful in donor countries might be worse than useless in different places and circumstances [17].*

In the USA, the institutions created for development projects overseas and at home, like the Office of Appropriate Technology (OAT) and the National Center for Appropriate Technology (NCAT), started rolling out technologies like solar energy, low-head hydroelectricity from abandoned dams, wind energy, recycling and composting, which provided alternate development path termed AT or one of its many pseudonyms [17]. There are a variety of pseudonyms used to describe AT including *progressive*, *alternative*, *light-capital*, *labour-intensive*, *indigenous*, *appropriate*, *low-cost*, *community*, *soft*, *radical*, *liberatory*, and *convivial* technology [18, 19]. Applications in the US addressed growing skepticism about the "role of technology in American life," thus engendering a debate that possessed significant cultural significance and ideological purpose but presented tangible and financial obstacles to established societal interests [17]. The economist E.F. Schumacher provided a framework for AT by categorizing it as *intermediate technology* that bridged the gap between indigenous technology and "capital-intensive technology of modern industry" [17, 18, 20]. Although apt in some regards, this characterization stigmatized AT as low-tech [21], which has proven incorrect with the recent proliferation of appropriate technologies like solar energy in all societies.

Appropriate technologies in their many forms and name have been traced back to the reign of Mao Tse-tung when it was used in rural small-scale industrialization while large-scale, capital-intensive industrialized technology was used in urban centers. In India, Mohandas Karamchand Gandhi, popularly referred to as the first appropriate technologist, shared a similar vision, "any concern with goods requires mass production, but concern with people necessitates production by the masses" [18]. His ideology inspired a group of economists in later years to found the Appropriate Technology Association of India [18].

Successes with AT in industrialized countries have yet to transfer to developing or low-income countries because conventional technical aid and development strategies operate under the notion that economic growth through industrialization equates to development [18]. This is evident in the MSWM sector, where waste-to-energy (WTE) and other advanced technologies are promoted in locations without local maintenance capacity [22]. However, recent discussions surrounding MSWM have raised concerns regarding the effectiveness of directly introducing these technologies into developing nations, considering their potential incompatibility of such regions [22].

Modern technology imposes dependence on replacement parts and linked technology, which is termed "the indivisibility of modern technology" [16]. A reliance that these low-income regions cannot afford. The capital drain on economies with tight budgets and the benefit to a minority segment of the population rather than the majority necessitates the consideration of alternative technologies in MWSM. "It is unreasonable not to promote appropriate technology for development in the traditional and informal sectors in view of the capital and foreign exchange situation in many Third World societies. Development in these regions must start with less complex and expensive techniques and move forward" [18]. At its core. AT is "context- and situation-specific," minimizing "financial, energy, transportation, and management costs and services" while "tackling community development problems" [18, 21]. Appropriate technology is not traditional technology. Rather it encompasses technologies that are suited to the socio-cultural milieu of the society involved [16].

### 1.2 Local context first: Contrasting MSWM requirements in high- and low-income settings

The current determinants for using technology and sustainable waste management in high-income countries are environmental protection, climate change, diminished natural resources, and public health. On the other hand, low-income countries facing stagnated or decreasing economic growth, "unequal distribution of wealth among people, socio-cultural limitations, local and international influences, and lack of effective national policies" understandably do not share the same sense of urgency or commitment to technological solutions to MSWM [23].

Thus, "Technology should be developed in keeping with local conditions" [3, 14]. The "what," "who," and "how" questions remain essential to the successful adaptation and longevity of current technology [1]. The financial capability, environmental assimilative capacity, and local priorities of a country or location should be considered in the determination of the suitable MSWM system and technology [14]. Therefore, local conditions and priorities should be the final determinant of the technology developed and implemented. "Context stands for the level on which a certain technology can effectively be organized" [16]. In this regard, the location of the proposed technology determines how well the technology can be implemented, managed, and utilized. For example, the availability of waste compositional and volume data make waste-to-energy technology a viable solid waste solution in high-income communities and countries, which is not necessarily the case in low-income countries which lack reliable waste production and characterization data [23]. Technological solutions must incorporate such considerations, thus necessitating a different approach in the northern and southern hemispheres.

In relation to the local context, or local considerations, the criteria in consideration of the application of technology in MSWM should include waste volume, waste stream, affordability, operability, regulations, acceptability, environmental concerns and land availability [3]. These criteria will, therefore, be used in this section to contrast waste management strategies between high- and low-income settings. Although the discussion resolves around countries, it should be noted that high-income countries still comprise low-income communities that could benefit from solutions derived for low-income countries. Similarly, low-income countries consist of high-income communities that benefit from modern MSWM technologies. Furthermore, the presence of "large" open dumps on all continents, Africa, Asia, Europe, South America, North America and the Caribbean [23] allude to the necessity of appropriate technology in all locations across the globe.

### *1.2.1 Waste volume and waste streams*

To a large extent, the economic status of a country determines the importance and corresponding complexity of the implemented MSWM systems. "The degree of attention paid to [MSWM system] sustainability varies from country to country and is correlated with economic status" [14]. Additionally, the economic status of a country is a strong indicator of the volume of waste generated, waste composition, and the complexity of the technology deployed [3, 4, 14, 23]. For example, the critical monitoring of waste generation increases with accuracy depending on the technology deployed. Accuracy in these measurements consequently enhances the effectiveness of the process design and the technology deployed in waste collection, transfer, treatment, and disposal. Waste production monitoring technologies developed to enhance waste collection include "geographic information systems (GISs), radio-frequency identification (RFID), ultrasonic sensors, and international system for mobile/general radio packet service (GSM/GPRS)" [23]. These technologies are available and suitable to high-income countries that can afford them while retaining the technological know-how for their operations and maintenance. However, these technologies are prohibitive in low-income countries because "untrained rag-picking is the sole method used for waste collection/segregation in about 63% of countries" [23]. This "improvised form of house-to-house collection involves a worker with a handcart who traverses each street. He rings a bell so residents can hear him coming, whereupon they leave their residences and deposit the waste in his cart. Once the handcart is full, the worker either unloads it in a community bin or deposits it in a transport vehicle" [14]. This working solution to a local waste collection challenge will prove ineffective in determining critical waste characterization and volume information. The system will provide data on transported waste but not necessarily the quantity of waste generated in the locale.

Waste generation rates, another technological determinant, vary in high- and low-income countries. For example, in 2016, low- and lower-middle-income countries contributed five and 29% of globally generated waste volumes. In contrast, upper-middle and high-income countries contributed 66% [4]. Therefore, strategies and permissible risks in the selection will significantly differ in urgency and acceptability. However, the projection that waste generation volumes in low-and lower-middle-income countries will surpass high-income countries by 2050 [4] should keep solid waste management a global priority.

Economic status is also a determining factor of the solid waste stream. Low-income countries generate more organic waste and thus will benefit more from biological technologies. Conversely, high-income countries have high recyclables with less organic waste. The low recyclable fraction is attributed to cultural and traditional practices, "In developing economies, recycling occurs at every stage of the system, leaving only a small portion that ultimately reaches the landfill for disposal" [14].

### *1.2.2 Affordability*

A switch to biological treatment will conserve energy use and result in significant financial savings due to its associated costs and unavailability in low-income countries. "The requirement for energy is an indispensable part of modern waste-treatment facilities. According to one estimate, about 2–3% of the energy consumed by a developing country is used to treat waste. One significant remedy to this problem is changing the mode of waste-treatment from mechanical to biological" [23]. However, in high-income countries, energy use in waste management is perversive and is only conserved due to environmental protection and conservation rather than scarcity.

While labour costs are a deterrent in high-income countries, this differs in low-income countries with abundant labour resources. Approximately 30% of solid waste management costs in a treatment plant are directed to salary and maintenance [23]. Automation in these plants has helped high-income countries reduce these costs. However, low-income countries need to benefit from their people resource. Rising service and compensation costs are a concern in high-income countries but not in low-income countries. Therefore, this requires differing and unique solutions for either location's circumstances. While the existing technologies offer benefits such as the provision of steam and electricity their long-term adverse effects on the environment, particularly in terms of greenhouse gas (GHG) emissions, diminish their appeal. Additionally, these technologies are financially demanding and necessitate infrastructure, thus limiting their feasibility, in low- and middle-income countries [23].

Capital investment in the substitution of equipment for labour thus becomes an economically justifiable action for high-income (developed) countries but not for low-income (developing and under-developed) countries.

### *1.2.3 Operability*

Operability in the context of this discussion is synonymous with operational efficiency. Operability necessitates the safe and reliable function of the entire MSW system. Therefore, the collection, transport, transfer, recovery, treatment, and disposal of the waste needs to be efficient and reliable. For example, the use of standardized collection

| Technological consideration | Attribute | High-income | Low-income |
|---|---|---|---|
| Waste volume | Rate of waste generation | High | Low |
| | Availability and reliability of waste generation/production data | Available and reliable | Unavailable and unreliable |
| | Projected waste volumes | Reduction per capita | Increase per capita |
| Waste stream | Waste composition and characteristics | High recyclables and lower biowaste. Low density. | Higher biowaste and lower recyclables. Low calorific value. |
| Affordability | Capital equipment substituted for labour | Yes. Prompted by rising service and compensation costs | No. Informal sector dependent, hence labour dependent |
| Operability | MSW system design | Long-term planning. Efficient and stable | Experimental |
| | Recycling sector | Well-organized collection and processing | Complex less-organized processes by small businesses and the informal sector |
| | Technological sophistication | Complex | Simple |
| | Knowledge of the technology | Trained and certified workers are available | Unavailable trained and certified MSWM administrators |
| | Energy | Available | Irregular and scarce |
| Acceptability | Priority of MSWM by governments and populace | High | Low |
| | Public engagement and participation | Satisfactory. Medium to high | Limited. Low |
| Land area availability | Land available for waste disposal | Low | Low |
| Environmental factors | Isolate landfill contents from the environment | Expensive preventative technology in use | |
| Regulatory constraints | Constraints in technological applications | Strict with strong enforcement | Weak enforcement [2] |

**Table 1.**
*Comparing technological considerations in high and low-income countries.*

vehicles and procedures that are consistent with the waste characteristics and weather patterns in Calgary, Canada, had the city designing a waste management system that staggered the organic waste collection from a weekly schedule in the summer to bi-weekly in the winter and spring seasons. The cooler temperatures are used to optimize the system's efficiency. These operational efficiency strategies are not seen in low-income countries that transport waste with a range of vehicles, from general-purpose trucks to highly mechanized compactors, which are difficult to maintain and inconsistent with the characteristics of the waste collected [14]. Attempts are being made in some locations to develop affordable collection vehicles better suited to local requirements [24]. Other design considerations that enhance operability include the availability of trained and certified workers and a MSW system design. The MSW system is designed for the long-term in high-income countries but short-term in low-income countries.

Finally, recycling in both economies have diverging approaches. Organized and cost-effective processes that leverage collaboration between public and private institutions and facilitate efficient systems are practiced in high-income countries [14]. On the other hand, the informal sector in low-income countries recovers recyclable resources from households, communities, and landfills, which are recycled through collaboration with private businesses [3, 24]. The informal sector is crucial in enhancing recycling rates, particularly in low to middle-income developing countries. The informal waste pickers collect waste from open dumpsites and streets, focusing on gathering recyclable materials and selling the valuable materials to formal or informal collection points. Consequently directly contributing to material recovery and the mitigation of environmental pollution [25].

Although well-suited to their locations, the efficiency of the informal waste recycling sector differs significantly from recycling in high-income countries, and it is difficult to quantify due to the lack of waste generation data, as mentioned earlier.

The differing technological criteria and their repute in low- and high-income settings are summarized in **Table 1**.

Organic waste is the most substantial solid waste fraction generated in municipalities worldwide. As such, their management is essential. Organic waste, biowaste, constitute 44% of waste volumes globally [4], which means that a substantial volume of waste can be diverted and recycled by targeting this type of waste. The Black Soldier Fly (BSF) treatment, one of the many biowaste treatment methods, is an economically viable biowaste management technology due to the by-products generated [26]. Furthermore, insect-based bioconversion is a burgeoning industry that produces larvae for human and animal feed and frass biofertilizer [27]. Hence the Black Soldier Fly is introduced as a suitable biological waste treatment alternative in both high- and low-income countries.

## 2. BSF technology, how it all began and why the growing interest?

The Black Soldier Fly (BSF), *Hermetia illucens*, is a non-pestiferous "large wasp-like fly" with exceptional biowaste management potential [28–30]. They are ravenous consumers of decaying biowaste, biodigesting double their body weight and thus recording a 50–80% biowaste volume reduction [29, 31, 32]. BSF waste treatment method is an emerging biowaste management technology used to valorize organic waste into frass biofertilizer while generating larvae for animal or human feed [27]. Studies of their application in waste management began in the 1970s but swelled at the turn of the century, with some authors citing it as a burgeoning technology [30, 33]. The economically viable by-products that BSF bioconversion produces make the technology very attractive. Mature Black Soldier Fly Larvae (BSFL), a product of the technology, contain 36–48% crude protein, 29–35% crude fat, amino acids, vitamins, and micronutrients, which makes them excellent as animal feed [28, 29, 34–36]. Entomology in biowaste management, i.e., insect farming with organic waste using BSF biotechnology, also fills the increasing demand for insect protein [35]. Various insects, including mealworms (*Tenebrio molitor, Alphitobius diaperinus*, and *Zophobas morio*), locusts (such as *Locusta migratoria* and *Schistocerca gregaria*), crickets (such as *Acheta domestica* and *Gryllodes sigillatus*), the house fly (*Musca domestica*), and BSF (*Hermetia illucens*), have been recognized as important protein contributors. However, BSF has gained notable recognition due to its remarkable ability to thrive on a wide range of organic waste materials [37].

BSF frass, the other biowaste product, is a processing residue used as a soil amendment or fertilizer [32]. The substitution of BSF frass for conventional fertilizers could reduce the global warming potential [38]. BSF are native to tropical and subtropical regions but have been noted in some temperate regions [34]. Being native to tropical and sub-tropical regions means the optimal environmental conditions for the entire BSF lifecycle are 25–32 degrees Celsius and 60% relative humidity [26]. Thus, climatic controls are required for BSF bioconversion in sub-optimal environmental conditions. This means its use in colder climates will require technological interventions that

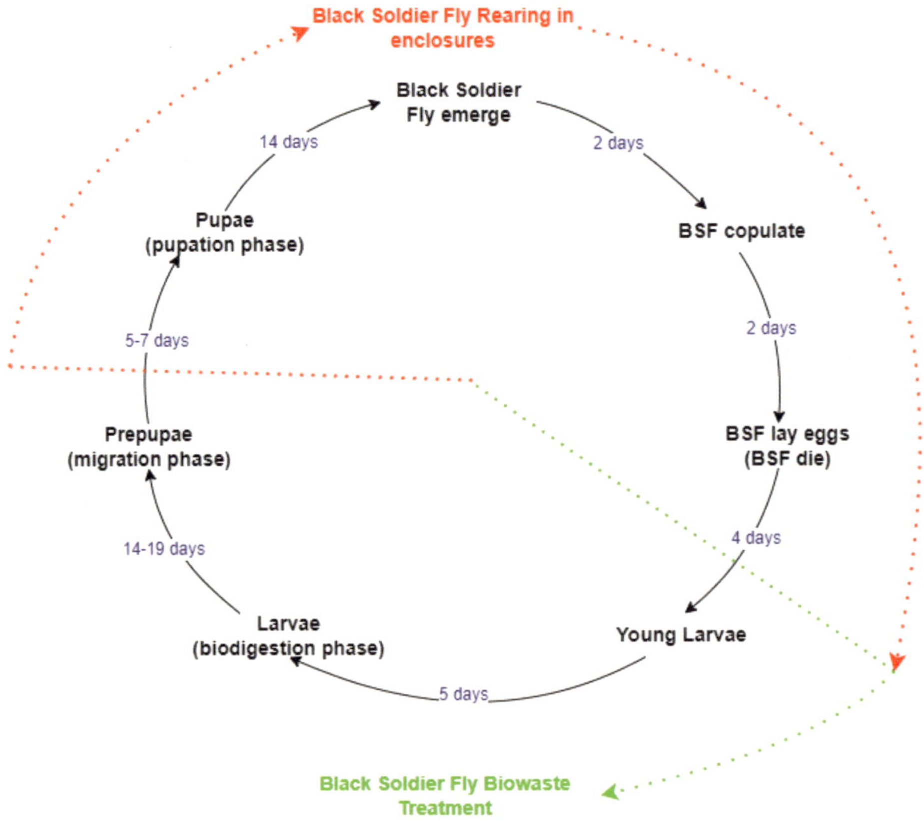


**Figure 1.**
*BSF waste treatment method in relation to BSF lifecycle.*

are optional in tropical climates. This makes BSF biotechnology an apt reference in analyzing location and context-specific technical application considerations.

The lifespan of a BSF begins at the egg phase (eggs laid), and it ends once the adult fly copulates and lays eggs (BSF dies) [39]. Depending on the BSF diet and environmental conditions, the time frame for the cycle from eggs, larvae, prepupae, pupae, and flies ranges between 20 and 44 days [26, 36, 37, 40, 41]. Adult flies live for about five to eight days [42], making the larvae and prepupae the most prolonged time interval for the BSF in the cycle. This feature proves advantageous in applying BSF in biotechnology as the most extended time interval of the BSFL is when they feed on biowaste. **Figure 1** represents the physiological and waste management lifecycles of BSF. When viewed as a biowaste management life cycle, the cycle can be subdivided into two; BSF breeding and BSF biowaste recycling. BSF Biowaste recycling includes prepupae harvesting and frass composting. These processes are external to the BSF life cycle but deliver the economic benefits of the BSF biological waste treatment method.

## 3. BSF solutions in high- and low-income settings

Several techniques have been employed in BSF biowaste management globally. They can be categorized into two groups, closed BSF systems based on the rearing of captive BSF and open systems, which depend on the natural oviposition of wild BSF [37]. Open systems are best-suited for low- and middle-income countries with tropical climatic conditions. In these locations, small-scale farmers, communities, and households can implement simplified units ranging from plastic trays to concrete trays and buckets [30, 37]. Although appropriate for the earlier cited application, this approach is "inefficient... to reach the desired production levels for meeting the livestock, poultry, and aquaculture feeds demand... colonization is unpredictable, resulting in lower production of BSF biomass" [37]. Therefore, the closed BSF system will be better suited for locations focusing on insect farming, i.e., BSFL as animal feed, aquaculture feed, or biofuel from the BSF fat and oil. The emphasis on BSF biomass is to obtain the maximum biowaste bioconversion. The closed system relies on the uninterrupted supply of BSF eggs from a rearing unit which is used to facilitate bioconversion with "somewhat predictable input (organic waste) and output (BSF biomass production)" [37]. Like the open system, this approach is implemented on various scales using different techniques. The main disadvantage of the closed system is the spatial requirement. Dortman et al. suggest 50 $m^2$ for the nursery and 100 $m^2$ for bioconversion per ton of incoming waste per day as the appropriate sizing of the BSF waste processing facility [39]. Large-scale facilities that treat 200 tonnes of waste/day for BSFL protein exist. However, information is commercially sensitive and not shared [34]. Such facilities require temperature and humidity control, automation and stable electricity and water supply, which implies increased costs available to high-income settings and the private sector. Numerous private companies are active in the production of BSF larvae. However, details about how they operate and their financial aspects are not made public. This lack of disclosure may be attributed to their desire to safeguard their competitive advantages [37]. However, one Costa Rican BSF treatment facility has been recorded as bioconverting 3 tonnes of municipal biowaste per day at a 930 $m^2$ facility producing 3.2 kg compost per $m^2$ per day [34].

The open system BSF biowaste treatment and biomass production are suitable for low-income communities because of their low capital investment and economic viability. Furthermore, the proposed BSF could also be sustainable, i.e., long-lasting, by integrating AT requirements. A case study, discussed in the next section, was developed by implementing an open system BSF treatment unit using AT techniques in a Dar es Salaam peri-urban community.

## 4. A community-based approach to the Black Soldier Fly waste treatment system: a Tanzanian case study

Affordability, technological simplicity, adaptability to inadequate infrastructure, and profitability are essential requirements of waste management systems in low-income countries [6, 22, 43, 44]. This case study investigates the appropriateness of the Black Soldier Fly (BSF) treatment in a Tanzanian community. The intent of the study was to co-design a self-sustaining biowaste management system that advanced the environmental and personal well-being of the community in which it was installed. The location-specific appropriateness of the BSF waste treatment method was analyzed using co-production methods in the system design, generation and utilization of the BSF larvae as animal feed and in using BSF frass for farming.

### 4.1 Integrating appropriate technology principles in the case study

The case study approach was based on the characterization that "Appropriate technology means simply any technology that makes the most economical use of a country's natural resources and its relative proportions of capital, labor and skills, and that furthers national and social goals" [18]. Therefore, the following criteria were employed in the design, implementation, and modification of the BSF study,

1. Context- and location-specific,
2. Considerate of the culture and economy,
3. Meet and satisfy the elementary needs of the community,
4. Stakeholder involvement in all phases of system design and operational planning and execution,
5. Minimal "financial, energy, transportation, and management costs and services"
6. No adverse environmental impact,
7. Adaptable and flexible to the community's needs,
8. Simplified technology that is "understood, controlled and maintained without high levels of education and training."
9. Small-scale and affordable, and
10. Facilitates development [21].

The investigation was subdivided into three phases to facilitate incorporating these requirements in the case study while optimally designing the BSF waste treatment.

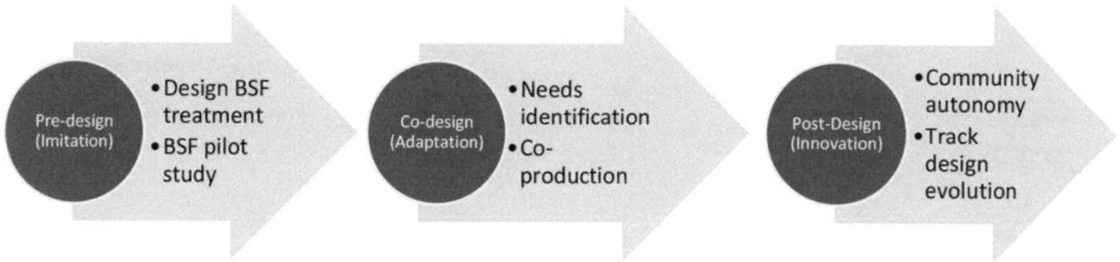


**Figure 2.**
*Appropriate technology system design in consideration of stakeholder needs, culture, economy and local context.*

The phases correlate with co-production strategies in creating innovative public services [45]. The three phases, *imitation*, *adaptation*, and *innovation,* were integrated with the project design stages, as shown in **Figure 2** above.

### 4.2 Pre-design and imitation

The pre-design phase of the study was used to conduct a review of the literature to determine the regional biowaste challenges, BSF development, and key actors in the BSF waste treatment method. Based on this background study, the key design parameters for the proposed case study included a co-designed BSF bioreactor, community-led project, adequate land space for the BSF facility, predictable, consistent, and adequate source of fresh biowaste, easy access and transportation to and from the waste source, non-toxicity of the waste, BSFL, and produced frass, zero or minimal energy and water requirement, and minimum maintenance.

The researcher also conducted a pilot BSF study to gain familiarity with the technology. A design brief highlighting the particulars of the case study was produced at the end of the pre-design phase. The brief included 3D renderings of the proposed BSF facility.

### 4.3 Co-design and adaptation

A partnership was established with AMREF Health Tanzania, an international public health non-governmental organization (NGO) with an extensive history of sponsoring and promoting waste-to-wealth initiatives in rural and urban areas in Tanzania. AMREFs water sanitation and hygiene (WASH) manager arranged formal introductions between the researchers and the municipality officials. The meetings were opportunities to gain an understanding of the location-specific goals and requirements of the municipality. The community and waste-to-wealth community group needs were captured through an AMREF-led focus group discussion and a formal introductory meeting. The initial formal meeting with the community group came first. AMREF chaired the session introducing the research team and research objective. The BSF waste treatment technology was then described in detail using the 3D renderings from the pre-design phase to describe the project better. The group then provided suggestions for and their expectations of the project. At the end of the meeting, a site for the BSF bioreactor was determined, and agreements were made on the elementary needs that should be satisfied, including minimizing odor nuisance and long-standing project support.

Finally, the focus group discussion was used to obtain the community's needs. **Table 2** captures the researcher-based pre-design requirements and the municipal authorities and community group's co-design requirements. All three stages were critical in the implementation of AT.

| Pre-design requirements | Municipality requirements | Community group requirements | Community requirements |
|---|---|---|---|
| Co-designed BSF bioreactor. | | Entire group involvement in design, planning, and operation. | |
| Community-led project (and a project leader). | | Community-led with protracted technical support | |
| Adequate land space for the BSF facility | | | |
| A predictable, consistent, and adequate source of fresh organic waste | | | |
| Easy access and transportation to and from the waste source | | | Easy access and transportation to and from the waste source |
| Non-toxicity of the waste, BSFL, and produced frass | Non-toxicity of the waste, BSFL, and produced frass | | |
| Zero or minimal energy and water requirement | | | |
| | Minimized odor nuisance | Minimized odor nuisance | |
| | Enables community development | Enables community development | Enables community cooperation |
| | | | Manageable with household responsibilities |

**Table 2.**
*Collating the final project requirement by integrating all stakeholder requirements.*

For the proposed BSF technology to truly be appropriate, the predetermined (pre-design) requirements needed modification by integrating the articulated needs of the stakeholders. The municipal authorities, community group, and community members in this instance. These needs also evolved as the project proceeded, but it was accommodated. This is the benefit of AT. A lesson learned in this instance was that although project requirements may be based on AT principles and generated from baseline studies (conducted by AMREF Health Tanzania) with extensive literature review, they can still differ from the stakeholder's priorities.

The co-design phase also included the research team's observations of BSF entrepreneurial activities in Dar es Salaam, visits to BSF facilities run by other organizations, discussions with waste researchers and entrepreneurs in the city, and biowaste characteristic and production pattern observations.

To satisfy the "Stakeholder involvement in all phases of system design and operational planning and execution" requirement of AT, co-design and co-production strategies were employed. Including key stakeholders in project management was encouraged and supported; co-creation and co-production made collaboration possible between the research team and the users. Co-production "is also known as co-creating services, whereby service recipients are involved in different stages of the process, including planning, design, delivery and audit of a public service" [46] and "co-creation refers to the active involvement of end-users in various stages of the production process" [47]. Both concepts focus on expanded end-user involvement as a means of incorporating as many requirements as possible and producing a product or service that is socially, economically, and environmentally appropriate, hence sustainable. "The concept co-creation and co-production seems to be related or maybe even interchangeable" [47] and as such, they will be used synonymously in this study.

"Co-production is more than mere consultations since it involves citizens/users in more systematic exchanges to create and deliver public services. Co-production transforms the relationship between service users and providers, ensuring greater user influence and ownership" [45]. The need for consultation in co-production quickly raised and resolved the community group's desire to have all their members involved through the entire project co-design phase. This choice made them more than end-users, and they became project participators. Their active and enthusiastic commitment saw the emergence of context- and location-specific innovative solutions, which will be detailed in the post-design section.

The first nine criteria of AT prompted the redesign of the original BSF facility. A shift from a closed BSF facility to an open system was made after consultation with representatives from AMREF Health, local tradesmen (construction), BioBuu Limited, a BSF business located in Dar es Salaam, and the community group. The AT criteria for simplified technology with minimal costs made the switch apparent and the decision easy.

The original BSF facility adopted the commonly practiced closed system with a BSF rearing section (nursery) and a bioconversion section [31, 34, 39, 48, 49]. **Figure 1** (closed system) and **Figure 3** (open system) demonstrate how both systems are designed around the BSF life cycle and thus differ.

However, as noted by AT practitioners, the application of homogeneous common practice has not effectively resolved the waste treatment challenge in low-income communities. So, the pre-designed system, **Figures 4** and **5**, was modified.

The revised co-created and co-produced BSF bioreactor, **Figures 6** and **7**, harnessed the resources naturally available to the group and location, like biowaste and wild BSF. Using wild BSF eliminated the need for a nursery, thus simplifying the BSF waste treatment system. Some elements of the pre-designed requirements were carried over, like a water channel around the bioreactor to deter predators, a leachate collection system to minimize the environmental impact of the treatment system, and chicken wire to exclude larger avian organisms.

Finally, *adaptation* in the co-design was palpable through the development and execution of the operational plan. The community group nominated a project leader based on the research team's recommendation of a *project champion*. The group traditionally elected a leader to new projects, which was a point of synergy between the research team and the community group. The project leaders, a woman and a man, were present at the BSF bioreactor during all the visits, and they ensured that information was communicated daily to the larger community group.

The consensus from the introductory meeting, focus group discussions, scoping sessions, and information from the pre-design literature review resulted in developing an operational plan that collected biowaste from surrounding households. This strategy proved unsuccessful because homes in the location practised direct-use waste management. This is the feeding of biowaste to household livestock. As such, the waste volumes collected for the first month of the project were minimal. An optimal solution was derived by engaging with the community group using co-production

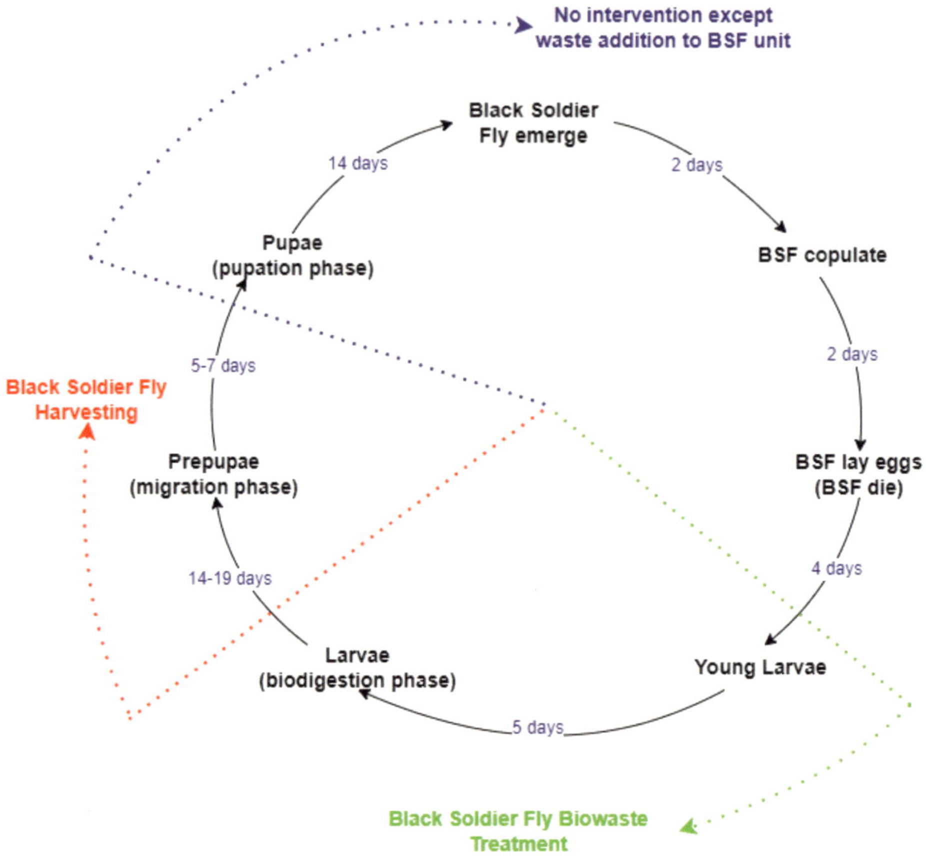


**Figure 3.**
*Open system BSF waste treatment method in relation to the BSF lifecycle.*

**Figure 4.**
*Pre-designed BSF rendering-Side view.*

**Figure 5.**
*Pre-designed BSF rendering-Aerial view.*

**Figure 6.**
*The co-created BSF bioreactor.*

strategies. This involved gathering fruit and vegetable waste from retail vendors at a nearby market, transporting the waste via tricycle to the BSF bioreactor, **Figure 8**, and adding the waste to the bin, **Figure 9**. The vendors typically had to pay for the waste collected once a week for transport to a landfill by the municipal authorities. By collecting the leftovers from the vendors just before midnight, at the end of the market day, the vendors eliminated waste collection fees, and biowaste was diverted from the landfill. Waste collection volumes increased from 41 kg/week to 365 kg/week in the first month. Thereafter, due to the enthusiasm and encouragement of the

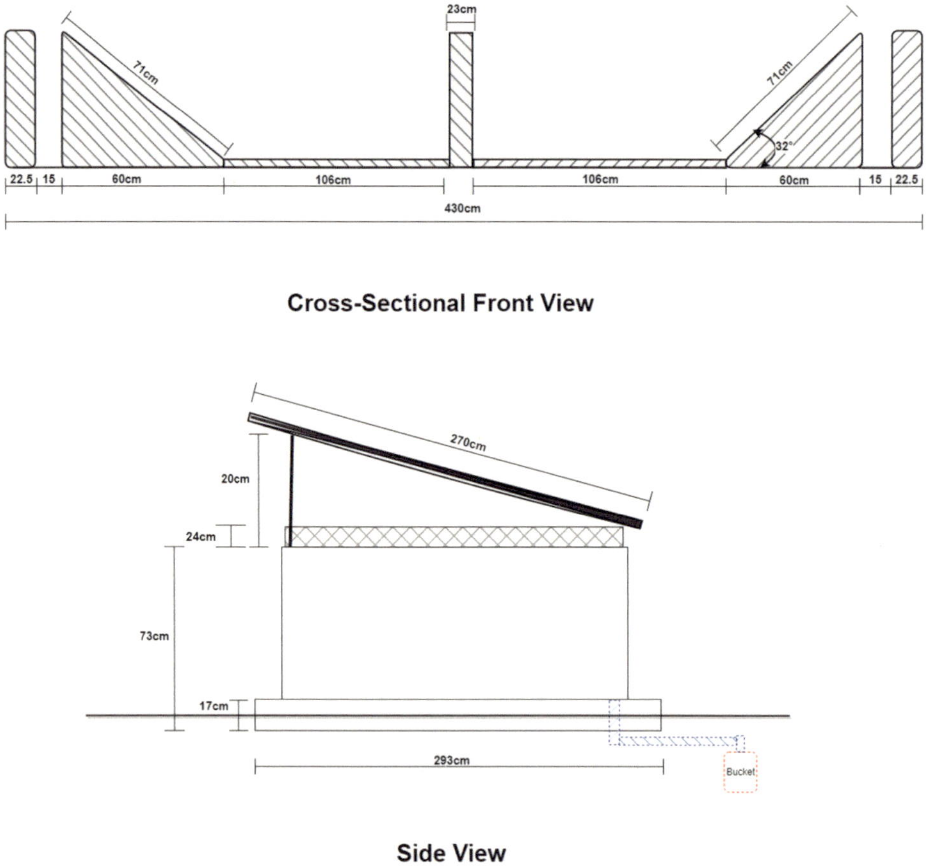


**Figure 7.**
*Case study BSF bioreactor schematics.*

**Figure 8.**
*Waste collected from the retail vendors is transported with a tricycle and weighed.*

community group, waste collection volumes increased to 460 kg/week. Other aspects of the project where co-production strategies were used to accomplish AT include the marketing and sales of the Black Soldier Fly Larvae (BSFL) and frass. As a result, the

**Figure 9.**
*The biowaste is accumulated for three weeks in alternating sections of the bioreactor.*

community became the largest customer of the products, and a member of the community group who is a farmer is currently the only beneficiary of the BSF frass.

### 4.4 Post-design and innovation

The post-design phase purposed two-fold objectives; the autonomous management of the BSF bioreactor by the community group and the identification of technological and operational improvements to the pre-designed system. Six cycles of BSF waste treatment were run between October 2022 and April 2023, and improvements were observed with the increasing number of cycles. The procedures followed for

**Figure 10.**
*Neighboring farmland. BSF bioreactor is located with the white building in the background on the right.*

| Appropriate technology criteria [21] | Cycle 1 (Oct–Nov) | Cycle 2 (Nov–Dec) | Cycle 3 (Dec–Jan) | Cycle 4 (Jan–Feb) | Cycle 5 (Feb–Mar) | Cycle 6 (Mar–Apr) |
|---|---|---|---|---|---|---|
| 1. Context- and location-specific | Co-design BSF unit | Shift to market waste | | Co-produced leachate collection system | | |
| 2. Considerate of culture and economy | | Appropriate permission was requested and granted from families of the women collecting waste at nighttime | | | | |
| 3. Meet and satisfy the elementary needs of the community | BSFL used by the community | Frass used on farms | | | | |
| 4. Stakeholder involvement in all phases | Stakeholders involved. | | | Stakeholder semi-autonomous | Stakeholder autonomous | |
| 5. Minimal costs | Yes, from cycle to cycle. | | | | | |
| 6. No adverse environmental impact | Yes, across all cycles. | | | | | |
| 7. Adaptable and flexible to community needs | | Waste collection volumes dictated by community | | Waste collection frequency is controlled by the community | | |
| 8. Simplified technology | BSF unit run with community | | | BSF unit run by the community alone | Frass drying unit co-produced | |
| 9. Small-scale and affordable | Yes | | | | | |
| 10. Facilitates the development | BSF project leaders trained | Community group leaders trained other members on the BSF system | | | Community designing BSF training centre. Interviewed on national | |

**Table 3.**
*Appropriate technology categorized improvements by BSF cycle.*

**Figure 11.**
*Trench outlet valve designed and implemented by the community group.*

each cycle differed, with each cycle becoming an improved version of the prior cycle. For example, cycles 3 and 4 differed significantly because the community gained semi-autonomous control of the operations and was in complete control by cycle 5. The research team strictly visited the site during these cycles for observation and data collection. In addition, the frass by-product became the highest-demand product from the BSF system. A productive BSF frass Chinese cabbage harvest resulted in the farmer, a community group leader, expanding the cultivated frass-applied land from 2 m × 6 m to 6.8 m × 12 m. The BSF bioreactor is located close to community farmland, **Figure 10** and farmers in the neighboring area have requested frass.

The lessons learned, and improvements made over each cycle are summarized in **Table 3** and categorized by AT criteria.

Modifications suggested and implemented by the community group in the innovation phase also include the addition of an outlet valve for cleaning soapy water in the trench surrounding the BSF bioreactor. The channel was built to deter rodents and predators that feed on the BSFL. Observations of BSF projects in the city revealed predators as an existent threat. Therefore, the water-filled trench was necessary, but surfactants were needed to prevent a hospitable breeding ground for mosquitoes. However, the weather conditions made algae thrive in the channel. Weekly operations, therefore, included manually cleaning the trench, which was arduous. The community group proposed installing a valve, **Figure 11**, which simplified weekly maintenance and increased the project viability.

Other innovations by the community group include a co-designed drying unit for wet harvested frass, weather- and rising ground water-proof leachate collection system, and BSF knowledge-sharing activities.

## 5. Conclusion

Waste management is as complex as populations are diverse. Solutions vary by region, income, season, and available technology. There exists a plethora of theoretical and practiced waste management solutions. However, no sustainable—economically, socially, and environmentally acceptable solution has emerged for low-income settings. Communities and countries with limited purchasing power have relied on the same technologies available to higher-income locations, resulting in abandoned MSW projects, persistent open dumping, and open burning.

Appropriate Technology has helped create a framework that enables technological application but with the careful long-term consideration of the stakeholder's culture, finances, environment, and social requirements. The Black Soldier Fly waste technology system is an ideal candidate for AT in low-income communities. The technology is used in various parts of the world at diverse scales, and it is flexible and adaptable enough to accommodate all AT requirements. It was successfully applied in a peri-urban Tanzanian location by members of a community group. The study is ongoing, but interest in BSF waste treatment is continuously growing in East Africa. The animal feed by-product (BSFL) and soil conditioner have proven successful in the case study and many others.

Food insecurity is a familiar concern for the future. However, people in low-income communities have raised their livestock for centuries, and BSF waste treatment allows communities to continue enjoying their traditions for centuries to come. AT with BSF waste treatment system is not a blanket solution to biowaste treatment for all low-income communities in every instance. But the combination hopefully encourages practitioners to consider all technological solutions, specifically the less advanced ones, because they might be the best-suited solution for the application. Finally, having multiple stakeholder categories involved in a project is arduous. Therefore, implementing AT criteria with co-design and co-production requirements using simple yet adaptable technology is a formidable challenge. Replicating technology becomes a more attractive option in the face of this daunting proposition. However, the longevity and productivity of the AT-produced technology more than make up for the time spent generating the lasting solution.

## Acknowledgements

This research was funded by the SOCIAL SCIENCES AND HUMANITIES RESEARCH COUNCIL, grant number 752-2022-1768 and Bioreactor Construction was funded by THE SCHOOL OF ARCHITECTURE, LANDSCAPE, AND PLANNING at THE UNIVERSITY OF CALGARY by the Research Expenses Award. The results published in this article is possible due to collaboration with AMREF Health Tanzania, Biobuu Limited, The International Institute of Tropical Agriculture (IITA) Tanzania, and Green Composting Limited.

## Conflict of interest

The authors declare no conflict of interest.

## Notes/thanks/other declarations

We are thankful for the support of the many who made this research possible including Dillion Consulting, SWANA Northern Lights Chapter, Air & Waste

Management Association (A&WMA), the members of Sauti ya Jamii Kipunguni, Grace Kisetu of Hema Homes, Dr. Steve Mbuligwe of Arhdi University, Dr. Mike Yhedgo, and Aliceanna Shoo, Mturi James, Anthony Ndjovu, Dr. Frida Ngalesoni, and Jane Tesha of AMREF Health Tanzania.

## Author details

Atinuke Chineme[1]*, Marwa Shumo[2], Getachew Assefa[1], Irene Herremans[1] and Barry Wylant[1]

1 University of Calgary, Calgary, Canada

2 Center for Development Research (ZEF), University of Bonn, Bonn, Germany

*Address all correspondence to: atinuke.chineme@ucalgary.ca

## References


[1] Schübeler P. Conceptual Framework for Municipal Solid Waste Management in Low-Income Countries. St. Gallen, Switzerland: SKAT (Swiss Centre for Development Cooperation in Technology and Management); 1996

[2] Jayasinghe R, Mushtaq U, Smythe TA, Baillie C. The Garbage Crisis: A Global Challenge for Engineers [Internet]. San Rafael, California (USA): Morgan & Claypool Publishers; 2013. p. 157. (Synthesis Lectures on Engineers, Technology, and Society). Available from: https://www-morganclaypool-com.ezproxy.lib.ucalgary.ca/doi/pdfplus/10.2200/S00453ED1V01Y201301ETS018 [Accessed: February 17, 2021]

[3] Iyamu HO, Anda M, Ho G. A review of municipal solid waste management in the BRIC and high-income countries: A thematic framework for low-income countries. Habitat International. 2020;**95**:102097

[4] Kaza S, Yao L, Bhada-Tata P, Van Woerden F. What a Waste 2.0: A Global Snapshot of Solid Waste Management to 2050 [Internet]. Washington, DC: The World Bank; 2018. p. 295. (What a Waste). Report No: 2. Available from: https://datatopics.worldbank.org/what-a-waste/trends_in_solid_waste_management.html [Accessed: November 19, 2020]

[5] Antwi-Agyei P, Dougill AJ, Agyekum TP, Stringer LC. Alignment between nationally determined contributions and the sustainable development goals for West Africa. Climate Policy. 2018;**18**(10):1296-1312

[6] Shava S, O'Donoghue R. Temporal cycles of waste management in Southern African indigenous societies. In: The Temporalities of Waste: Out of Sight, Out of Time. London, United Kingdom: Routledge; 2020

[7] Hendriksen A, Tukahirwa J, Oosterveer PJM, Arthur PJ. Participatory decision making for sanitation improvements in unplanned urban settlements in East Africa. Journal of Environment & Development. 2011;**21**(1):98-119

[8] Kazaure MB. Survey on SWM for sustainable development and public health in Dutse Metropolis, Jigawa state, Nigeria. Procedia Environmental Sciences. 2016;**35**:57-64

[9] Ezedike C, Ohazurike E, Emetumah FC, Ajaegbu OO. Healthseeking behavior and waste management practices among women in major urban markets in Owerri, Nigeria. AIMS Public Health. 2020;**7**(1):169-187

[10] Ajibade LT. Indigenous knowledge system of waste management in Nigeria. IJTK. 2007;**6**(4):642-647

[11] Gani BA, Chiroma A, Gana BA. Women and solid waste segregation in Bauchi Nigeria. Journal of Environment and Earth Science. 2012;**2**(8):22

[12] Kosoe EA, Diawuo F, Osumanu IK. Looking into the past: Rethinking traditional ways of solid waste management in the Jaman South Municipality, Ghana. Ghana Journal of Geography. 2019;**11**(1):228-244

[13] Siragusa L, Arzyutov D. Nothing goes to waste: Sustainable practices of re-use among Indigenous groups in the Russian North. Current Opinion

in Environmental Sustainability. 2020;**43**:41-48

[14] Shekdar AV. Sustainable solid waste management: An integrated approach for Asian countries. Waste Management. 2009;**29**(4):1438-1448

[15] Tharakan J. Indigenous knowledge systems for appropriate technology development. Indigenous People. 2017;**123**:123-134

[16] Endresen SB, Hesselberg J. The concept "appropriate technology" and development in the third world. Norsk Geografisk Tidsskrift-Norwegian Journal of Geography. 1987;**41**(3):151-154

[17] Pursell C. The rise and fall of the appropriate technology movement in the United States, 1965-1985. Technology and Culture. 1993;**34**(3):629-637

[18] Akubue A. Appropriate technology for socioeconomic development in third world countries. JOTS [Internet]; 2000;**26**(1):33-48. Available from: https://scholar.lib.vt.edu/ejournals/JOTS/Winter-Spring-2000/akabue.html [Accessed: April 26, 2023]

[19] Haynes KE, El-Hakim SM. Appropriate technology and public policy: The urban waste management system in Cairo. Geographical Review. 1979;**69**(1):101-108

[20] Donaldson K. The future of design for development: Three questions. Information Technologies and International Development. 2009;**5**(4):97-100

[21] Tharakan J. Educating engineers in appropriate technology for development. World Transactions on Engineering and Technology Education. 2006;**5**(1):233-235

[22] Oteng-Ababio M, Melara Arguello JE, Gabbay O. Solid waste management in African cities: Sorting the facts from the fads in Accra, Ghana. Habitat International. 2013;**39**:96-104

[23] Das S, Lee SH, Kumar P, Kim KH, Lee SS, Bhattacharya SS. Solid waste management: Scope and the challenge of sustainability. Journal of Cleaner Production. 2019;**228**:658-678

[24] Medina M. Solid Wastes, Poverty and the Environment in Developing Country Cities: Challenges and Opportunities. Helsinki: WIDER; 2010. Available from: http://hdl.handle.net/10419/54107 [Accessed: December 8, 2020]

[25] Ferronato N, Portillo MAG, Lizarazu GEG, Torretta V. Formal and informal waste selective collection in developing megacities: Analysis of residents' involvement in Bolivia. Waste Management & Research. 2020;**39**(1):108-121

[26] Lohri CR, Diener S, Zabaleta I, Mertenat A, Zurbrügg C. Treatment technologies for urban solid biowaste to create value products: A review with focus on low- and middle-income settings. Reviews in Environmental Science and Biotechnology. 2017;**16**(1):81-130

[27] Ojha S, Bußler S, Schlüter OK. Food waste valorisation and circular economy concepts in insect production and processing. Waste Management. 2020;**118**:600-609

[28] Zhang J, Huang L, He J, Tomberlin JK, Li J, Lei C, et al. An artificial light source influences mating and oviposition of black soldier flies, Hermetia illucens. Journal of Insect Science [Internet]. 2010;**10**(1):1-7. DOI: 10.1673/031.010.20201

[29] Singh A, Kumari K. An inclusive approach for organic waste treatment and valorisation using Black Soldier Fly larvae: A review. Journal of Environmental Management. 2019;**251**:109569

[30] Diener S, Gutiérrez FR, Zurbrügg C, Tockner K. Are larvae of the Black Soldier Fly–Hermatia illucens–a financially viable option for organic waste management in Costa Rica? In: Proceedings Sardinia 2009, Twelfth International Waste Management and Landfill Symposium. S. Margherita di Pula, Cagliari, Italy: CISA Publisher, Italy; 2009

[31] Zabaleta I, Mertenat A, Scholten L, Zurbrügg C. Selecting Organic Waste Treatment Technologies [Internet]. Dubendorf, Switzerland: Eawag-Swiss Federal Institute of Aquatic Science and Technology; 2020. Available from: https://www.eawag.ch/fileadmin/Domain1/Abteilungen/sandec/schwerpunkte/swm/SOWATT/sowatt.pdf [Accessed: October 7, 2021]

[32] Beesigamukama D, Mochoge B, Korir N, Musyoka MW, Fiaboe KKM, Nakimbugwe D, et al. Nitrogen fertilizer equivalence of Black Soldier Fly frass fertilizer and synchrony of nitrogen mineralization for maize production. Agronomy. 2020;**10**(9):1A151-1A151

[33] Quilliam RS, Nuku-Adeku C, Maquart P, Little D, Newton R, Murray F. Integrating insect frass biofertilisers into sustainable peri-urban agro-food systems. Journal of Insects as Food and Feed. 2020;**6**(3):315-322

[34] da Silva GDP, Hesselberg T. A review of the use of Black Soldier Fly larvae, Hermetia illucens (Diptera: Stratiomyidae), to compost organic waste in tropical regions. Neotropical Entomology. 2020;**49**(2):151-162

[35] Wang YS, Shelomi M. Review of Black Soldier Fly (Hermetia illucens) as animal feed and human food. Food. 2017;**6**(10):91

[36] Diener S, Zurbrügg C, Tockner K. Conversion of organic material by Black Soldier Fly larvae: Establishing optimal feeding rates. Waste Management & Research. 2009;**27**(6):603-610

[37] Surendra KC, Tomberlin JK, van Huis A, Cammack JA, Heckmann LHL, Khanal SK. Rethinking organic wastes bioconversion: Evaluating the potential of the Black Soldier Fly (Hermetia illucens (L.)) (Diptera: Stratiomyidae) (BSF). Waste Management. 2020;**117**:58-80

[38] Lopes IG, Yong JW, Lalander C. Frass derived from Black Soldier Fly larvae treatment of biodegradable wastes. A critical review and future perspectives. Waste Management. 2022;**142**:65-76

[39] Dortmans B, Diener S, Verstappen B, Zurbrügg C. Black Soldier Fly Biowaste Processing. A Step-by-Step Guide [Internet]. Dubendorf, Switzerland: Eawag–Swiss Federal Institute of Aquatic Science and Technology; 2017. Available from: https://www.eawag.ch/fileadmin/Domain1/Abteilungen/sandec/publikationen/SWM/BSF/BSF_Biowaste_Processing_HR.pdf [Accessed: July 15, 2021]

[40] Alvarez L. The role of black Soldier Fly, *Hermetia illucens* (L.) (Diptera: Stratiomyidae) in sustainable waste management in northern climates [Internet] [Ph.D]. Windsor: University of Windsor; 2012. Available from: https://scholar.uwindsor.ca/etd/402 [Accessed: July 12, 2021]

[41] Čičková H, Newton GL, Lacy RC, Kozánek M. The use of fly

larvae for organic waste treatment. Waste Management. 2015;**35**:68-80

[42] Mutafela RN. High value organic waste treatment via black Soldier Fly bioconversion [Master's]. Stockholm, Sweden: KTH Royal Institute of Technology; 2015

[43] Asomani-Boateng R. Closing the loop: Community-based organic solid waste recycling, urban gardening, and land use planning in Ghana, West Africa. Journal of Planning Education and Research. 2007;**27**(2):132-145

[44] Hammed T, Sridhar M, Olaseha I. Effect of demographic characteristics and perceptions of community residents on solid waste management practices in Orita-Aperin, Ibadan, Nigeria. Journal of Environmental Systems. 2011;**33**:187-199

[45] Pestoff V. Collective action and the sustainability of co-production. Public Management Review. 2014;**16**(3):383-401

[46] Realpe A, Wallace LM. What Is Co-Production? [Internet]. London, UK: Coventry University; 2010. p. 11. Available from: https://improve.bmj.com/sites/default/files/resources/what_is_co-production.pdf [Accessed: May 13, 2023]

[47] Voorberg WH, Bekkers VJJM, Tummers LG. A systematic review of co-creation and co-production: Embarking on the social innovation journey. Public Management Review. 2015;**17**(9):1333-1357

[48] Diener S, Studt Solano NM, Roa Gutiérrez F, Zurbrügg C, Tockner K. Biological treatment of municipal organic waste using Black Soldier Fly larvae. Waste and Biomass Valorization. 2011;**2**(4):357-363

[49] Caplin R. Business models in sustainability transitions: A study of organic waste management and black Soldier Fly production in Kenya [Internet] [Master's]. Oxford, United Kingdom: Smith School of Enterprise and the Environment University of Oxford; 2022. Available from: https://www.linkedin.com/posts/ryancaplin_business-models-for-organic-waste-and-bsf-activity-6988508871459766272--5KL/ [Accessed: November 22, 2022]

Chapter 7

# Characterization of Dismissed Landfills via Geophysical Techniques

*Lorenzo De Carlo, Giorgio Cassiani, Rita Deiana, Gian Piero Deidda and Maria Clementina Caputo*

## Abstract

In the context of waste landfill management, geophysical methods are a powerful tool for evaluating their impact on public health and environment. Noninvasive and cost-effective geophysical techniques rapidly investigate large areas with no impact on the system. This is essential for the characterization of the waste body and the evaluation of the liner integrity at the bottom of the landfill and leakage localization. Three case studies are described with the purpose of highlighting the potentiality of such techniques in landfill studies. The case studies show different site conditions (capped landfills, controlled closed systems, and unconfined systems) that limit the applicability of any other kind of investigation and, at the same time, highlight the versatility of the geophysical techniques to adapt to several field situations. Electrical and electromagnetic techniques proved to be the most efficient geophysical techniques for providing useful information to develop an accurate site conceptual model.

**Keywords:** landfill characterization, environmental protection, geophysical imaging, electrical resistivity tomography, electromagnetic induction, geophysical modeling

## 1. Introduction

The characterization of operational or dismissed municipal solid waste (MSW) landfills requires great care, given the potential negative impact of such sites on human health and their environmental exposure. Toxics such as PCBs organic compounds and heavy metals, leachate that accumulates as a result of the degradation of solid matrices and chemical reactions that take place within the waste body, and gas produced by the bacterial fermentation of organic waste in oxygen-free environments are critical elements to be accurately monitored. Before 1990s, landfills were designed without any technical requirements, usually filling dismissed quarries or abandoned areas filled with all kinds of waste. These uncontrolled systems caused several environmental disasters due to the spread of contaminants that, conveyed by rainfall, infiltrated through the vadose zone before reaching the groundwater. Over time, community regulations imposed strict constraints both in the landfill design and in the operational and postoperational plant management. In order to meet the necessary conditions for preventing pollution of the soil, subsoil, groundwater, or surface water, EU legislation (Council Directive 1999/31/EC) introduced appropriate technical measures consisting of the combination of a geological barrier and a bottom liner, generally made of high-density polyethylene (HDPE), during the operational stage and a top liner (capping) during the postclosure one. During the life cycle of the landfill, the main chemical-physical parameters, both of the liquid and gas phases, need to be monitored by means analytical measurements on samples extracted by collection network systems installed inside the landfill body in order to periodically verify they did not overcome warning thresholds. In addition, a network of warning wells located both upstream and downstream of the landfill is required for the monitoring of the qualitative status of the groundwater nearby of the landfill. However, such constrains do not guarantee real-time safety control of the landfill. Indeed, often potential rupture of the impermeable HDPE liner is detected only after the contamination reaches the underlying groundwater, requiring costly environmental remediation measures. More information about the integrity of the HDPE liner, as well as waste body hydrodynamics (moisture content, leachate accumulation, gas migration, etc.) are crucial for better understanding the landfills status, but direct investigations are not recommended due to the landfill safety risk connected. In light of these considerations, it is essential to identify noninvasive investigation techniques that can provide valuable information on the state of the thick waste layer and the mechanisms that take place inside it. This is even more important in case of capped landfill, i.e., when the impermeable liner has been laid on the ground surface in the postoperational stage to prevent infiltration of rainwater into the waste that would increase the leachate emission. In such context, in the last decades, geophysical techniques proved to be a powerful tool for characterization of landfills by providing detailed information otherwise undetectable with other indirect techniques. Geophysics is a noninvasive, cost-effective imaging technique of the Earth able to identify subsurface structures through measurements of physical parameters collected on the surface or in boreholes, strictly correlated petrophysical relationships to geological, hydrogeological and geotechnical properties of the investigated subsoil. A plentiful scientific literature reports detailed study cases where geophysical techniques have been successfully applied for landfill characterization. Many authors described electrical [1–7], electromagnetic [8], and seismic [9–11] investigations for general reconstruction of uncontrolled buried landfill areas or confined systems. In most cases, a combination of several geophysical techniques, such as Electrical Resistivity Tomography (ERT), Vertical Electrical Sounding (VES), Mise-a-la-Masse (MALM), Induced Polarization (IP), Self-Potential (SP), Electromagnetic Induction (EMI), Ground Penetrating Radar (GPR) and refraction seismic data allowed for reducing the uncertainty in the identification of the geometry of waste deposits, lateral extent of the waste, identification of the filling thickness, differentiation between surface cover and waste deposits, estimation of volumes of buried solid waste [12–17]. In case of closed landfill, i.e., when the impermeable HDPE liner is laid down on the bottom of the landfill, the evaluation of the liner integrity and leak detection has been successfully reported with potential measurements by using MALM technique [18–26], integrated MALM and ERT [27], GPR [28], and EMI [29]. Other study cases refer to the use of ERT [30–33], EMI [34–37], or a combination of several techniques, included GPR [38], IP and SP [39], and electrical and seismic data [40] for mapping of leachate accumulation zones. Such techniques have also been applied for monitoring the leachate flow [41–54], monitoring leachate injection and recirculation to enhance waste biodegradation [55–59], and identifying heavy metal sludge disposal [60, 61]. Finally, widespread literature reports geophysical applications for estimating water content [62–66], mapping biogeochemically active zones in waste deposits through IP [67], inferring subsurface gas dynamics and gas emissions [68, 69] and detecting groundwater contamination induced from leachate percolation [70, 71]. In this chapter three case studies, two located in the Apulia Region, south of Italy, and one in Veneto Region, North of Italy, are reported with the aim of enhancing the potentiality of the proposed geophysical techniques for landfill studies both in uncontrolled and confined closed systems. Different site conditions limited the applicability of any other investigation and, at the same time, highlighted the versatility of the geophysical techniques to adapt to several field features. Specifically, the three case studies concern:

- a dismissed MSW landfill investigated with an integrated approach of electrical techniques and numerical modeling;
- a capped MSW landfill, explored with electromagnetic measurements in order to overcome the strong limitation due to the capping on the top surface;
- a capped MSW landfill where electrical data, collected with an unconventional electrodes configuration along the perimeter of the landfill, have been integrated with numerical modeling.

## 2. Methods

The aim of this paragraph is to describe the electrical and the electromagnetic techniques, that are the most effective geophysical techniques used for landfill characterization, for several reasons. First, they are sensitive to the highly conductive structures due to the high organic, moisture, and leachate content of the waste deposits. In addition, the strong electrical contrast between the waste body, leachate, and the surrounding rock medium allows for a rapid identification of localized anomalies, in case of lateral or vertical spread beyond the landfill liner. Moreover, the impermeable HDPE liner represents a hydraulic barrier and, at the same time, an electrical insulator, the continuity of which can be easily checked with these techniques. Finally, these techniques can investigate large areas, such as those interested in landfills, with variable vertical and lateral spatial resolution.

### 2.1 Electrical methods

Electrical methods are based on measurements using stainless steel electrodes implanted into the ground. Four electrodes array is commonly used for collecting electrical resistivity or its inverse electrical conductivity. A couple of electrodes are used for injected DC current (**Figure 1A** and **B**), into the ground. The electric current flows depending on the subsurface electrical properties and the potential difference is measured by another pair of electrodes (M and N).

According to the Archie's law, the apparent resistivity $\rho_a$ ($\Omega$ m) of the ground, is determined (Eq. (1)):

$$\rho_a = \frac{K\Delta V}{I} \quad (1)$$

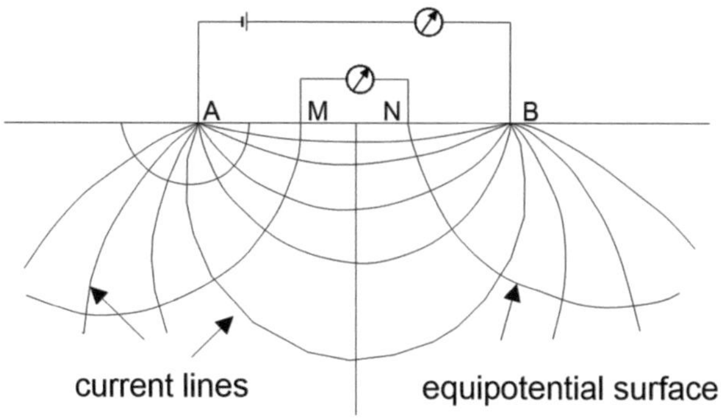


**Figure 1.**
*Distribution of electric current in the half-space.*

where ΔV is the difference potential (V), I is the injected current (A), and K is a geometric factor depending on the electrodes arrangement.

The measured value $\rho_a$ is an "apparent" resistivity because it refers to the resistivity of a homogeneous ground measured from a specific electrode arrangement. Apparent electrical resistivity is not a physical measurement, and the complex relationship with the "true" resistivity is obtained by solving the "inversion" problem.

Several electrical resistivity techniques are currently used with the quadrupole array. Here we focused on two different techniques: electrical resistivity tomography (ERT) and mise-a-la-masse (MALM).

### *2.1.1 Electrical resistivity tomography (ERT)*

Electrical Resistivity Tomography (ERT) is the most widely used electrical technique for the subsurface imaging. More details about ERT technique are described in [72].

This technique combines a high sequence of quadrupole measurements in order to obtain a dense grid of apparent resistivity data according to a tomographic scheme (**Figure 2**). The observed data provide a qualitative representation of the subsurface electrical properties by contouring the apparent resistivity data. The "real" resistivity is achieved from inversion, that is a numerical technique that tries to determine a subsurface resistivity model by minimizing a generalized objective function, i.e., the misfit between the calculated (theoretical value) and observed (experimental value) data at the nodes of a mesh into which the subsurface is divided.

The resolution, i.e., the minimum distance at which a target can be detected, depends mainly on the inter-electrode spacing used for data collection, while the depth of investigation is influenced by the length of the ERT profile, the distance between current and potential dipoles, as well as the electrical properties of the investigated medium.

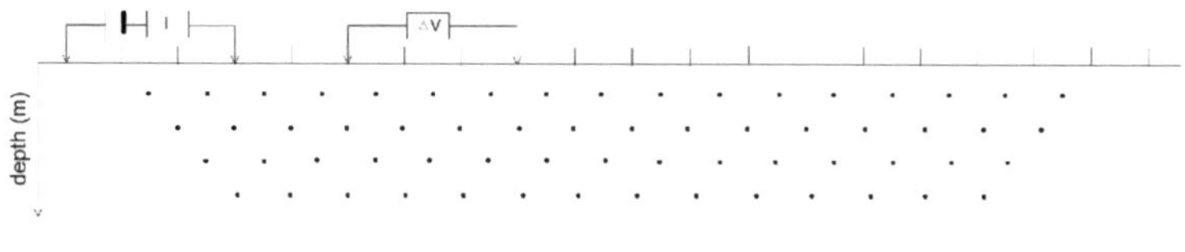


**Figure 2.**
*Distribution of apparent electrical resistivity collected by a quadrupole array.*

Depending on the data collection mode, along transects or in an area, the inverted model provides 2D or 3D resistivity model of the ground, which take into account lateral and vertical changes in the subsurface layers. Collecting resistivity data in time-lapse mode, i.e., repeating resistivity measurements over time along profiles or in an area, allows for monitoring spatiotemporal variations associated with dynamical processes.

#### *2.1.2 Mise-a-la-masse (MALM)*

Mise-a-la-masse is one of the oldest electrical resistivity techniques originally used to identify and map boundaries and orientation of electrically conductive ore bodies [73–76].

In MALM technique, the four-electrode array is set up as follows: two electric currents are placed (A) inside the conductive target and (B) ideally at infinity, other two potential electrodes are placed (N) outside at infinity opposite of B and (M) is used to detect the edges of the underground target (**Figure 3**).

The electric current flows from inside to outside the target, and a potential map is recorded with reference to the electrode M.

When used for detecting the integrity of the HDPE liner laid down on the bottom of the landfill, no current pathways are recorded between the inner and outer parts of the landfill and, consequently, a zero potential gradient is measured if the liner has no ruptures. In case of lack, the electrical current moves toward the infinity electrode (B) across the rupture of the liner and a voltage map indicate the presence of current pathways between the inner and outer parts of the landfill.

### 2.2 Electromagnetic induction (EMI)

The EMI technique is based on the electromagnetic induction principle. An alternating current circulates in a transmitter loop coil T, generating a time-varying magnetic field, defined as the primary magnetic field ($H_P$). This field induces eddy currents in a conductive body placed into the ground, which in turn generate a secondary magnetic field ($H_S$), having an amplitude much smaller than that of the primary field. A receiver coil (R) measures the total complex electromagnetic response, i.e., the sum of both primary and secondary magnetic fields (**Figure 4**).

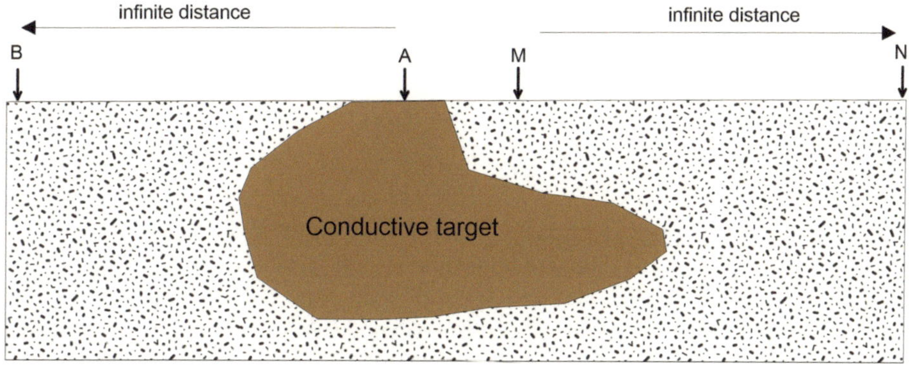


**Figure 3.**
*Basics of MALM technique.*

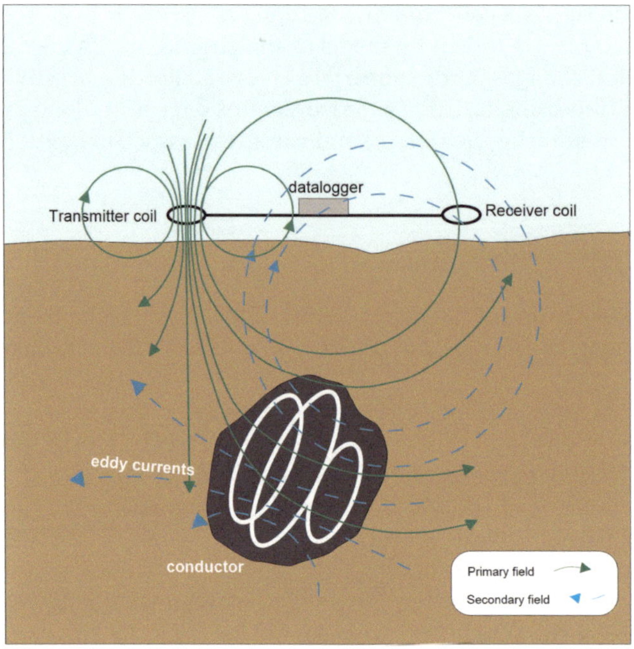


**Figure 4.**
*Basics of EMI technique.*

The real part, which has the same phase as the primary magnetic field, is called Inphase (P) component while the imaginary part, called Quadrature (Q) component, is 90° out-of-phase with the primary. The response parameter, called also "induction number," indicated with β, depends on electrical resistivity, the magnetic permeability, the geometry of the buried body, and ω, the operating angular frequency of the current in the transmitter coil (Eq. (2)):

$$\beta = \frac{r}{\delta} = \frac{r}{\sqrt{\frac{2}{\omega \mu_0 \sigma}}} \tag{2}$$

When β < <1, usually defined as "low induction number" (LIN) condition [77], the complex electromagnetic response can be simplified so that Q is linearly proportional to the electrical conductivity of the half-space $\sigma_a$ (Eq. (3)) and P is mainly affected by the magnetic permeability.

$$\sigma_a = \frac{4}{\mu_0 \omega r^2} \bullet \Im \left( \frac{H_S}{H_P} \right)_{\substack{HCP \\ VCP}} \tag{3}$$

The response parameter is a function of the penetration depth of the signal, defined as the skin depth δ, that is the depth at which an electromagnetic signal attenuates by a value equal to 37% with respect to the initial value. It depends not only by the electromagnetic properties of the investigated medium but also on the coils distance. At the same time the depth of investigation is also a function of the coils orientation. When they are oriented perpendicular to the ground, the horizontal magnetic dipoles are in vertical coplanar configuration (VMD), which allows the investigation of the superficial layers. By rotating both coils by 90°, the vertical magnetic dipoles turn into horizontal coplanar position (HMD), which increases the investigation depth. As for raw ERT measurements, the data measured is an "apparent" electrical conductivity (ECa), i.e., it is the equivalent electrical conductivity of a homogeneous half-space that produces the same measured response to the instrument in a single configuration (coil distance, coil orientation, frequency). In addition, the high electrical conductivity of the waste deposits, combined with the high inter-coil distance necessary for increasing the depth of investigation, makes it impossible that the LIN condition is met, causing a nonlinear EM response of the half-space and leading to a data inversion procedure for modeling the landfill body.

## 3. Case studies

### 3.1 Corigliano d'Otranto landfill

The first case study concerns a dismissed quarry, located in Southern Italy, near the municipality of Corigliano d'Otranto (4451117 N–265375.00 E WGS UTM84), used as MSW landfill since late 1980s. During the excavation of a new landfill, bordering the old one, leachate came out at the base of the scarp, which separates the two landfills (**Figure 5**), clear evidence of the leaching out of the old landfill. Few direct information was available in the original landfill design: the waste thickness was

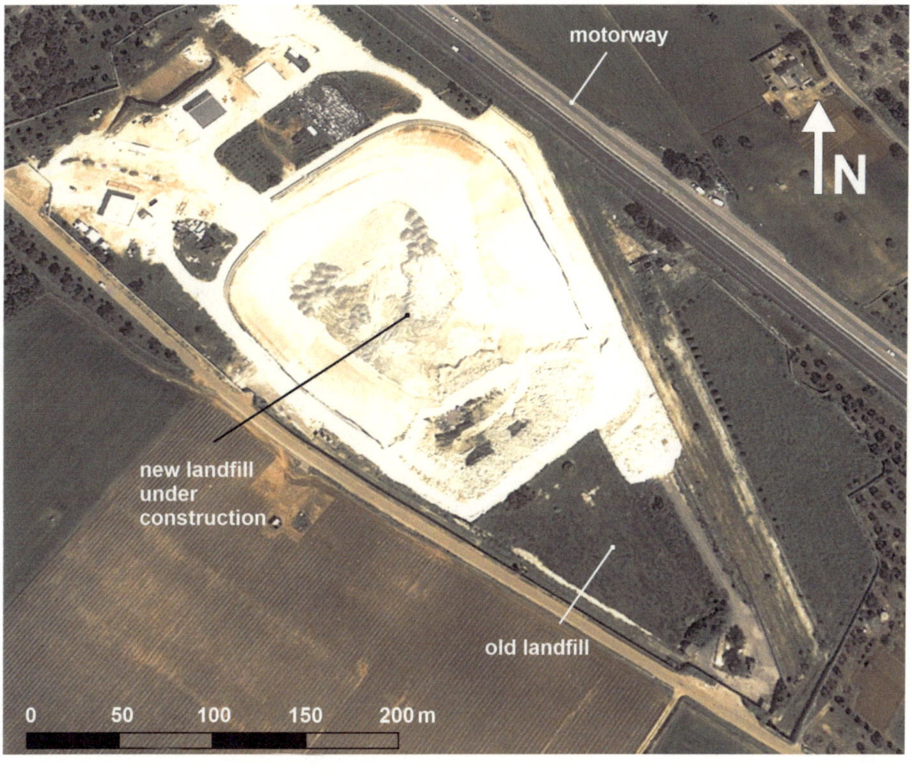


**Figure 5.**
*The landfill area at Corigliano d'Otranto.*

no greater than 20 m from ground surface, and the existence of a barrier on the landfill bottom, proved by the presence of the impermeable HDPE liner, clearly visible along the South-West border of the landfill. These uncertainties requested the planning of a detailed geophysical investigation in order to evaluate the liner integrity at the landfill bottom and localize possible ruptures and spill points.

For this purpose, ERT and MALM survey were carried out (**Figure 6**). In particular, two ERT profiles, named Line 1 and Line 2, were carried out inside the landfill in order to identify the approximate thickness of the waste in the old quarry. Line 1 runs along the dirt road separating the old and new landfills, Line 2, almost perpendicular to Line 1, crosses both the old and the new landfills along the scarp between the two landfills. In addition, another ERT profile, Line 3, was carried out outside the landfill as a reference section for defining the resistivity distribution of the rocky subsurface not interested in waste deposits. ERT data were collected using an IRIS Syscal Pro 48 (Iris Instruments) resistivity meter, by planting 48 stainless steel electrodes into the ground with 5 m spacing and 235 m long array. A Wenner–Schlumberger configuration was used because it offers a reasonable compromise between resolution and investigation depth, about 40 m from the ground surface. Overall, more than 2000 resistance quadrupoles were collected for each profile, including direct acquisition and reciprocal data, i.e., by swapping current with potential dipoles, in order to provide a correct data error estimation [78]. In the preprocessing stage, data filtering has consisted in removing measurements that exceeded 10% of reciprocal error.

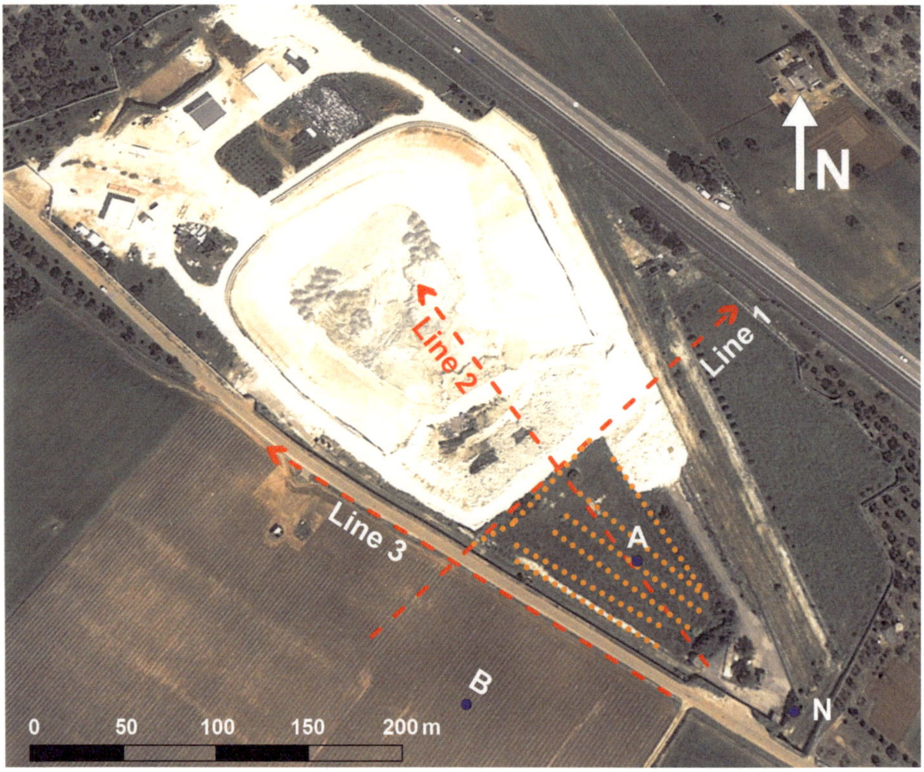


**Figure 6.**
*Location of the ERT and MALM measurements carried out at the Corigliano site. The red dashed lines represent the 2D ERT profiles, while the orange dots indicate the location of the potential electrodes used for the MALM survey.*

Inversion of ERT data was run using the code ProfileR (A. Binley—Lancaster University), an inverse solution for a 2-D resistivity distribution based on computation of 3-D current flow using a quadrilateral finite element mesh. In addition, seven MALM profiles were performed inside the old landfill, with 2 m inter-electrode spacing in order to ensure a good spatial coverage of the whole area. Electric current was injected between electrodes A and B, and voltage was measured with reference to electrode P. Potential data were mapped using Surfer (Golden Software, LLC) code.

### 3.2 Ugento landfill

The second case study refers the geophysical investigation performed in the MSW landfill located in Ugento (4420400 N–2795270 E WGS UTM84). The landfill was built in the 1990s in an abandoned quarry of calcarenite, a sedimentary porous rock widespread in the Apulia Region. It is divided into three plots and covers a total area of about 9 hectares (**Figure 7**).

The thickness of the waste body ranges from 15 m (plot I) to 19 m (plots II and III). An impermeable HDPE liner is laid down at the bottom and along the lateral boundary of each plot, as clearly visualized in the time-lapse historical Google Earth pictures of the landfill area. At the end of the operational stage, the Ugento landfill has been capped in order to prevent infiltration of rainwater into the waste that would increase the leachate production and gas dispersion in the air. The presence of the impermeable HDPE liner on the top surface inhibits any direct investigation and strongly limits the use of other indirect surveys. Being based on the electromagnetic induction principle, the EMI technique overcomes such strong limitations because it does not require any galvanic contact between probes and the ground. For this purpose, an EMI survey was planned using a CMD DUO device (GF Instruments), which consists of two independent coils, a transmitter, and a receiver, connected to each other by a flexible cable of different length, 10 m, 20 m, and 40 m, respectively, to deepen the electromagnetic signal into the waste deposits and the underlying vadose zone. Twenty-one EMI transects were collected either in a VMD or HMD configuration coil on the top of the landfill along parallel profiles in two orthogonal directions: 12 profiles in NW-SE direction, approximately 20 m apart, and 9 profiles in NE-SW direction, approximately 40 m apart (**Figure 8**). Continuous measurements mode, i.e. the data are collected while the instrument moves in the field, has been chosen by setting a time step equal to 2 s in order to obtain stable hence accurate measurements.

Overall, more than 11,000 measurements have been collected combining six different configurations, resulting in using three inter-coil spacings for each of the two coil configurations. Such data have been merged in order to obtain 1540 geometric depth soundings with six complex responses each, which, in turn, have been used for inversion. A free MATLAB software package, FDEMTool code [78], was used for inverting the electromagnetic data, and the resulting model was visualized with Voxler (Golden Software, LLC) code.

### 3.3 Asiago landfill

The MSW Asiago landfill (700550 N–5081863 E WGS UTM84) was developed since 2001 in a closed limestone quarry. The landfill has an area of about 1.5 hectares with a perimeter of about 500 m, and a maximum thickness of the waste deposits of about 40 m. The landfill was closed in mid-2018, and ever since it is a totally closed

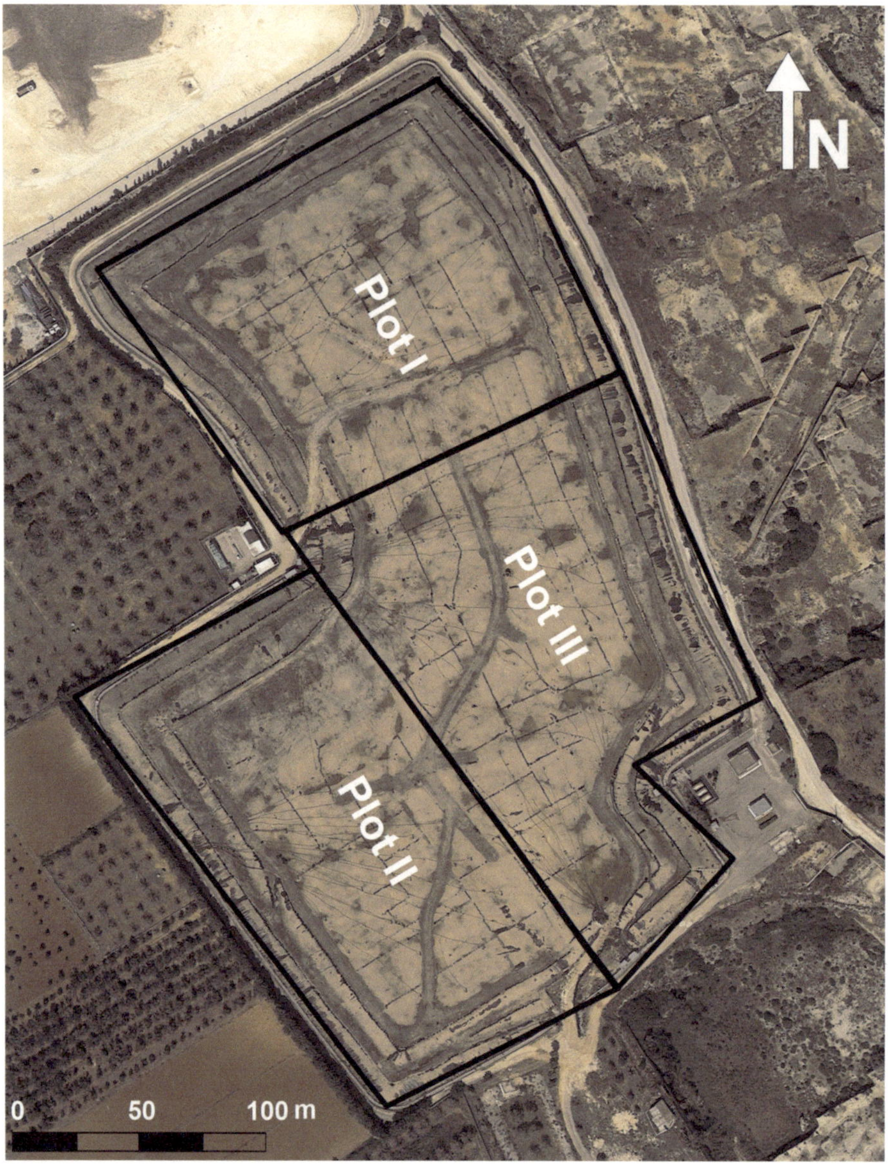


**Figure 7.**
*The landfill area at Ugento site (image source: Google Earth).*

system due to the presence of HDPE liners at its bottom and at its top surface (**Figure 9**).

Clear evidence of karstic phenomena reported in the area of the landfill site raised concern for the possible presence of large karstic cavities below the landfill and, secondarily, possible pathways for leachate contamination through the limestone fractured system. In search of possible large karstic cavities and possible leachate leaks, an ERT survey was proposed and implemented. As data collection from the top surface of the landfill body was made impossible by the presence of the impermeable (and electrically insulating) HDPE liner, an unconventional electrode configuration

**Figure 8.**
*Location of the EMI measurements carried out at the Ugento site. The black lines represent the 2D EMI transects (image source: Google Earth).*

has been chosen by placing 101 electrodes all around the perimeter of the landfill body (**Figure 10**).

A Syscal Pro (Iris Instruments) resistivimeter was used for the data collection. Overall, more than 12,000 resistivity measurements were collected, including reciprocals, to evaluate data quality. Data filtering rejected bad data points that overcome threshold values equal to 10% of reciprocal error and measured voltages less than $10^{-6}$ V.

**Figure 9.**
*The landfill area at Asiago site (image source: Google earth).*

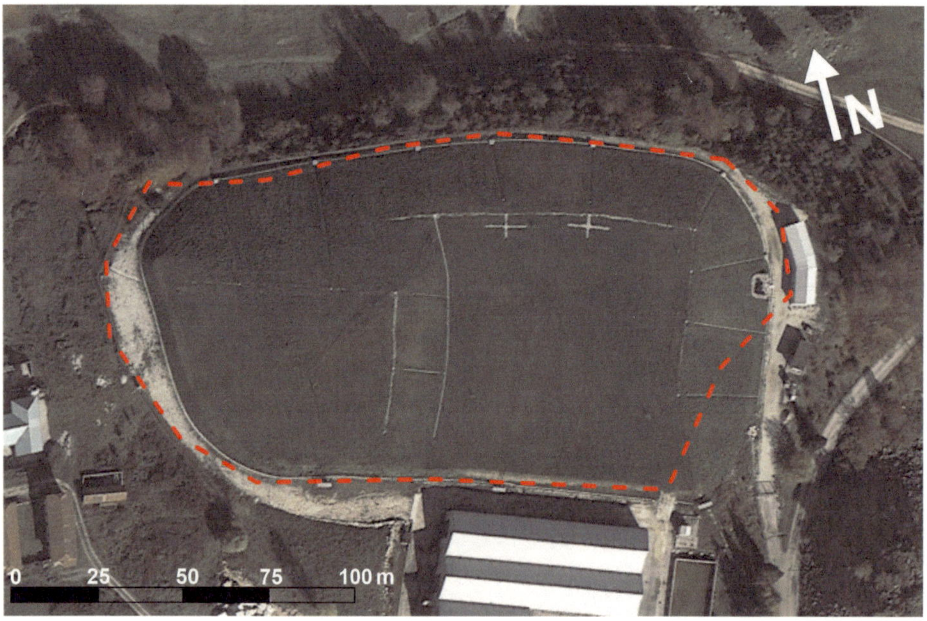


**Figure 10.**
*Location of 101 electrodes placed around the perimeter of the landfill body (dashed red line) at the Asiago landfill site (image source: Google Earth).*

The survey geometry was designed on the basis of preliminary synthetic modeling based on a 3D model of electrical current diffusion in a heterogeneous medium. The purpose was to verify whether the proposed approach was capable of imaging large karstic cavities (of a diameter around 20 m) and thus identify a viable acquisition and inversion strategy. Three scenarios, based on the presence of a cavity below the bottom of the landfill at different locations, were simulated. This posed the basis for a comparison with the results from real observed data.

## 4. Results

### 4.1 Corigliano d'Otranto landfill

The inverted ERT cross-sections clearly show the electrical behavior of the waste deposits and the rocky subsurface. Line 3, located outside the landfill and hence not involved in waste deposits, highlights a homogeneous distribution of the electrical resistivity, with a narrow range, expressed in logarithmic scale, varying from 2.6 and 3 Ω m, typically associated with the porous rock that constitutes the vadose subsurface

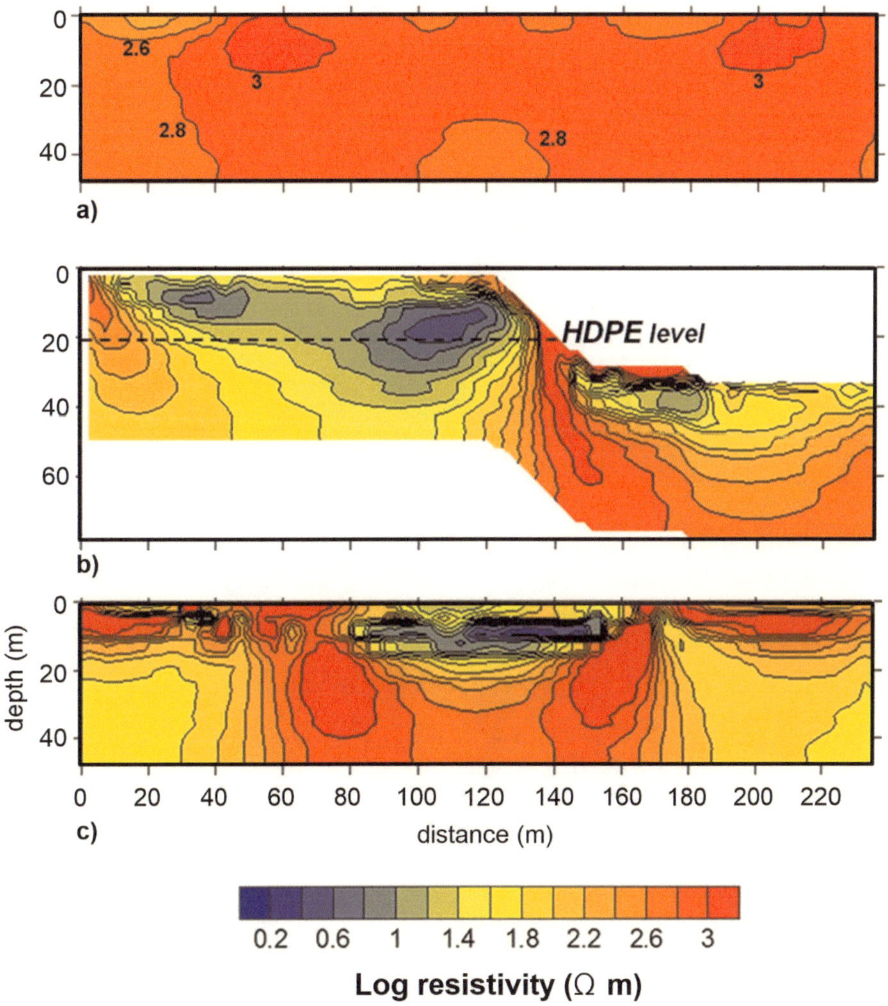


**Figure 11.**
*2D ERT cross-section from: (a) Line 3; (b) Line 2; (c) Line 1.*

(**Figure 11a**). On the contrary, Lines 2 and 1 (**Figure 11b** and **c**, respectively) put in evidence a very high conductive structure, with $\log_{10}$ resistivity approximately below 1.6 Ω m, clearly associated with the waste deposits located in the old landfill. Differences between the two ERT cross-sections are observed. In particular, while in Line 2, the conductive body is confined within the 20 m from top surface, Line 1, which cuts the scarp along its maximum slope, shows a conductive body deepening much more than the expected depth of the HDPE liner (dashed black line), about 20 m from the top surface. This difference can be explained by the fact that Line 1 violates the assumption, always made in 2D ERT, that the ground perpendicular to the survey line is homogeneous. In fact, the presence of waste deposits located in the old landfill on one side of the profile and the free space (air) on the other side of Line 1 masks the true thickness of the waste deposits.

On the other hand, this assumption is verified in Line 2, and the corresponding ERT cross-section provides a correct quantitative interpretation of the thickness of the waste deposits. The scenario derived from Line 2 indicates a vertical migration of conductive leachate which penetrates the supposed impermeable landfill bottom. The MALM provides further information about the structure of the landfill body. The potential map shows a sharp voltage jump in correspondence with the limits of the old landfill, in particular across the visible emergence of the HDPE liner along the South–West boundary (**Figure 12**).

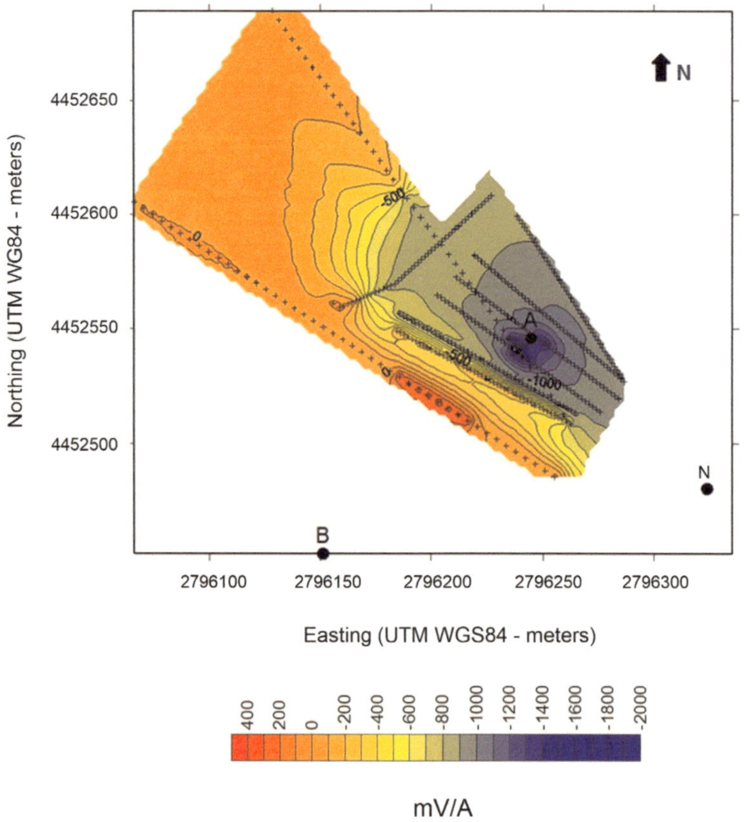


**Figure 12.**
*Potential map obtained from MALM survey.*

This evidence could be due to the poor confinement close to ground surface, considered that no other evidence of the liner is present on the ground surface. As the MALM results do not clearly support the ERT findings, in order to corroborate the hypothesis of the rupture of the impermeable HDPE liner, different alternative interpretation scenarios have been compared with numerical modeling. **Figure 13** shows the simulations of the geophysical modeling (ERT—Line 2 and MALM) in case of undamaged and damaged HDPE liner scenario, respectively. The simulated MALM maps do not show significant differences with the observed maps in both predicted scenarios (**Figure 13c** and **d** compared with **Figure 12**), probably due to the poor lateral confinement previously mentioned. On the other hand, the simulation of the ERT—Line 2 cross clearly highlights a good agreement in case of damaged HDPE liner scenario (**Figures 11b** and **13b**), thus leading to reasonably assume the hypothesis of a rupture of the HDPE liner through which the current flows.

## 4.2 Ugento landfill

The distribution of the electrical conductivity in the landfill body is visualized by a 3D image (**Figure 14**). It can be schematized into a three-layers model: (a) upper highly conductive; (b) weakly conductive intermediate; (c) resistive lower layer.

The visualization of the findings through 2D vertical cross sections (**Figure 15**) better highlights the upper conductive layer, with values from 1000 mS/m to 1800 mS/m, and the lower resistive layer, having lower values to about 150 mS/m.

Depth slices at different depths put in evidence details of the electrical conductivity inside the landfill (**Figure 16**).

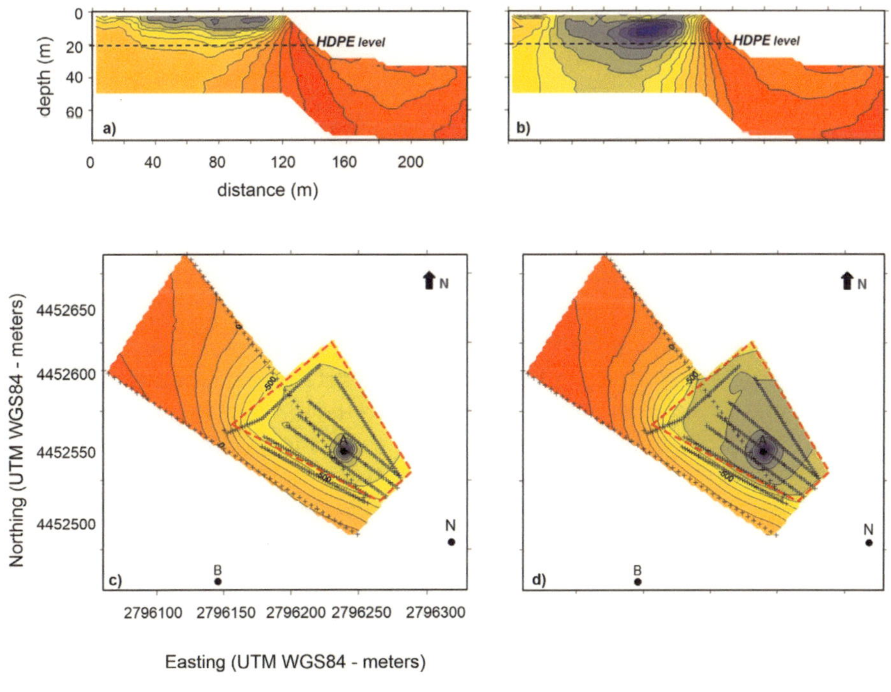


**Figure 13.**
*Simulated ERT—Line 2 cross-section and MALM potential map in case of undamaged HDPE liner scenario (**Figure 13a** and **c**) and damaged HDPE liner scenario (**Figure 13b** and **d**).*

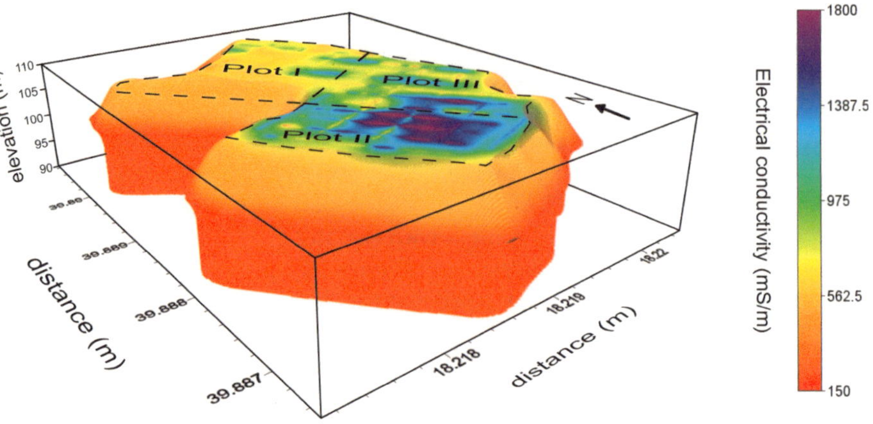


**Figure 14.**
*3D electrical conductivity model of Ugento landfill.*

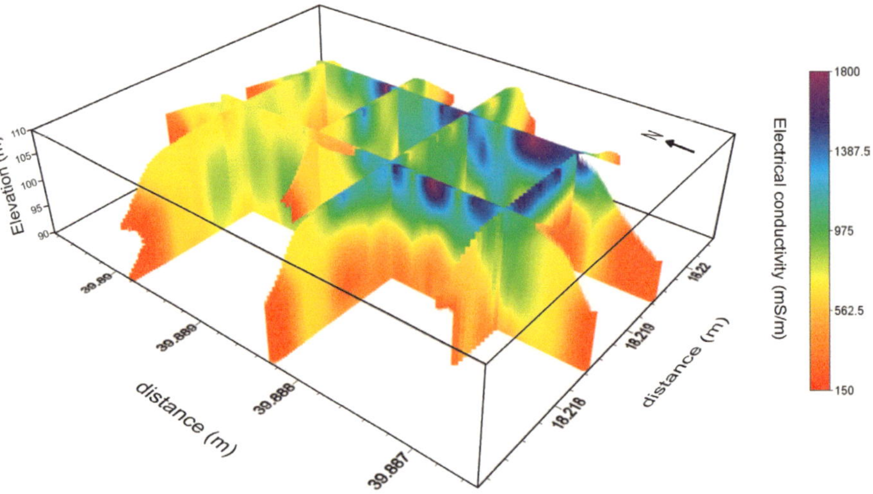


**Figure 15.**
*3D visualization of the electrical conductivity model through 2d vertical cross sections.*

The highly conductive upper layer has a maximum thickness of 5 m and an average EC ≈ 1200 mS/m (**Figure 16a** and **b**). The observed heterogeneity is probably associated with different waste composition, probably due to organic waste, a higher moisture content or a different compaction of the waste body by means of compacting trucks. The weakly conductive intermediate layer ranges from 5 m to 10 m from ground surface with an average value of EC ≈ 500 mS/m (**Figure 16c** and **d**). The waste deposits highlight a gradual decreasing in conductivity, explained by the loss of water in the waste body due to the weight of the overburden mass and then collected by the pipelines at the bottom of the landfill, and/or with the limited content of materials with high surface conductivity. Finally, the resistive lower layer (bedrock) deepens up to the maximum investigation depth with minimum EC ≈ 150 mS/m. In particular, no significant variations in the inverted electromagnetic signal were recorded in this portion of the subsurface (**Figure 16e** and **f**). Any expected high conductivity signal, linked to accumulation of leachate, is detected at depth, thus ensuring their absence.

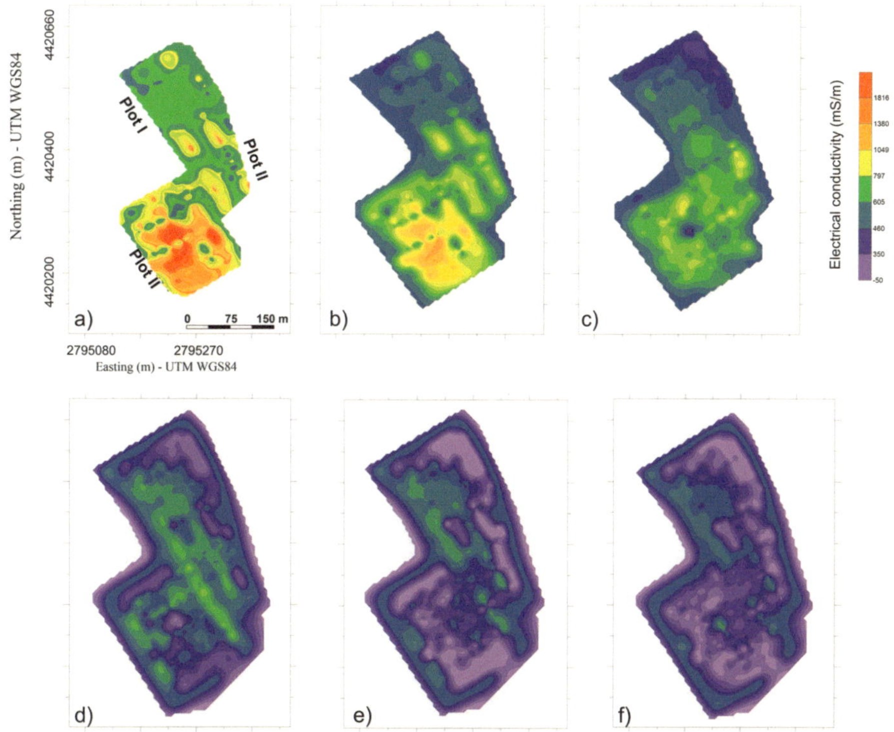


**Figure 16.**
*2D depth slices extracted from the 3D EC model: (a) z = 1 m from ground surface; (b) z = 2 m from ground surface; (c) z = 5 m; (d) z = 10 m; (e) z = 15 m; and (f) z = 20 m.*

## 4.3 Asiago landfill

The field acquisition at the Asiago landfill was conducted in one day (June 1, 2019) following the same configuration that proved most effective in the synthetic exercise, i.e., a total of 101 electrodes with a full dipole-dipole skip-12 configuration, that corresponds to a dipole length of 13 × 5 m = 65 m. **Figure 17** shows the results of the inversion of real field data. No significant high resistivity anomalies, associated with potential karstic cavities, have been detected, leading to exclude their presence. However, as a side result, the ERT survey showed the presence of a low-resistivity zone in the downhill portion of the landfill, as highlighted in the red circle: note that in this region, very close to the electrodes themselves, the sensitivity is particularly high. This area also corresponds to a low in the topography of the quarry where the landfill was developed. Since this high conductivity anomaly could be linked to the presence of leachate, a direct investigation was planned and performed in July 2021: the drilled core showed the presence of fine sediments (silt and clay) accumulated in this area of valley bottom, and the presence of freshwater probably coming from uphill in the valley, with no relationship with the landfill itself.

The real 3D resistivity image, compared against the expected image (**Figure 18**) in the presence of a 20 m diameter cavity, confirms that no such cavity exists below the landfill. Smaller cavities, if present, would have no substantial impact on the mechanical stability of the system. As the presence of karstic cavities was the main goal of the

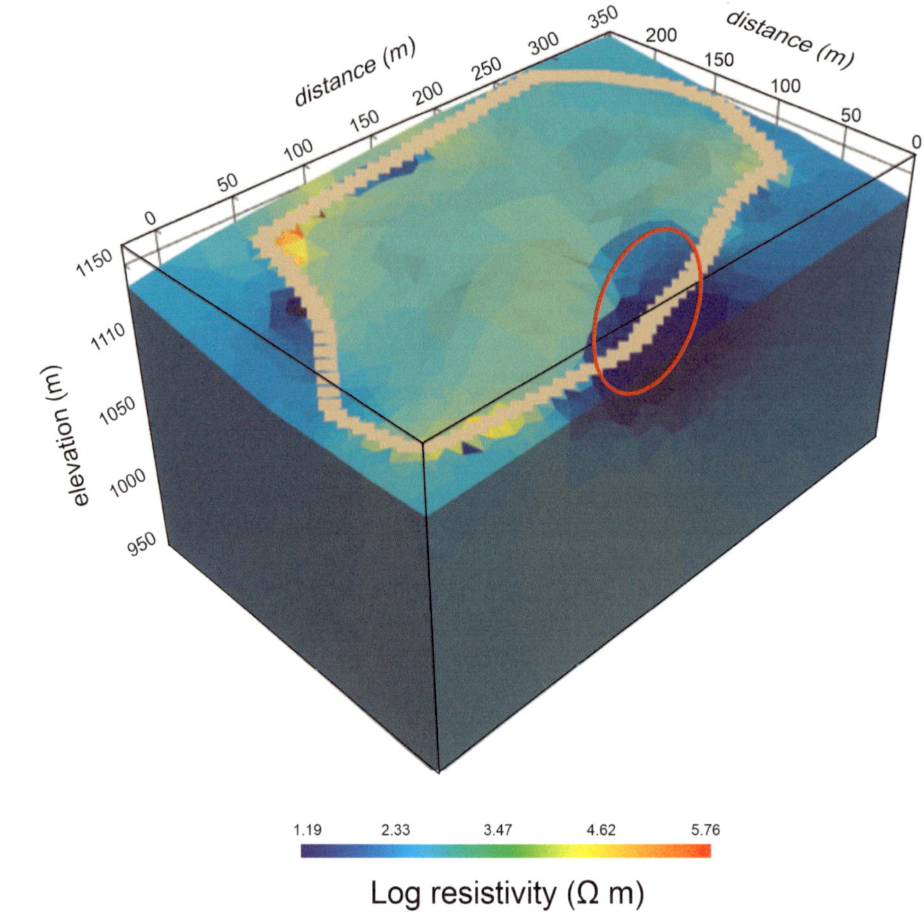


**Figure 17.**
*Inverted resistivity model, expressed in log scale.*

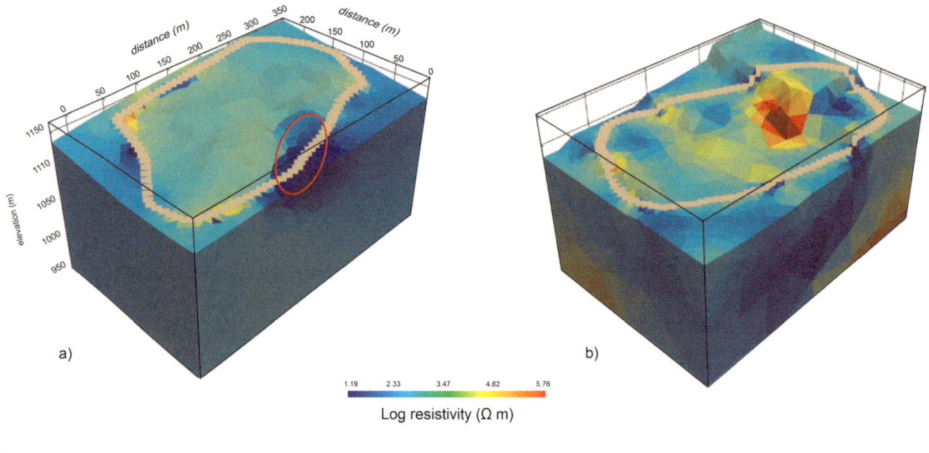


**Figure 18.**
*Comparison between inverted (a) and simulated (b) ERT model.*

investigation, one could conclude that the survey confirmed that no such feature is present at the site.

## 5. Conclusion

In the context of the study of abandoned or dismissed landfill, the lack of crucial information concerning landfill design and waste body characterization makes difficult the development of an accurate site conceptual model. Clear evidence about the presence of the HDPE impermeable liner on the bottom of the landfill and its possible integrity, as well as uncertainty on information regarding the structure of the waste deposits, are often missing. Noninvasive geophysical techniques have proved to be an effective tool to "see" inside the waste deposits in order to provide useful information about their physical properties with an accuracy degree unattainable with other technique. A huge amount of data can be collected over large areas in a short time in order to image the thickness of the covering layer, the geometry of the waste deposits, type of waste, and potential accumulation zones. In addition, data collected in time-lapse mode allows for monitoring hydrodynamic processes taking place inside landfill body. In this chapter, three case studies showing different site conditions have been described in order to highlight the potentialities and the versatility of the geophysical techniques in the study of landfills. In the Corigliano d'Otranto landfill, ERT and MALM measurements have been collected inside and outside the landfill to estimate the thickness of the waste deposits, verify the confinement of the waste deposits above the impermeable HDPE liner and, in case of rupture, image the leachate infiltration in the underlying vadose zone. The ERT outcomes highlight a very low-resistivity structure (less than 10 Ω m), associated to the waste deposits, one order of magnitude lower than the rocky subsurface. The thickness of this conductive structure is about 40 m, much more than the expected 20 m, leading to hypothesize a rupture of the impermeable HDPE liner and a vertical migration of landfill leachate through the landfill bottom. On the other hand, the potential map obtained from MALM measurements cannot clearly confirm such hypothesis, due to the lack of insulation in the shallow layer covering the waste mass which causes a current flow from inside to outside of the landfill. In such cases, when logistical limitations strongly affect the geophysical data acquisition and interpretation, forward numerical modeling provides added value to the geophysical data, by simulating different alternative interpretation scenarios. The good agreement between simulated and observed data in case of damaged HDPE liner leads to reasonably assume the hypothesis of a rupture of the bottom liner through which the current flows. In the capped Ugento landfill, the presence of the HDPE impermeable liner on the top surface strongly limits any direct investigation. The EMI measurements allowed to gain information about the waste composition, the thickness of the waste deposits and integrity of the HDPE liner. The electromagnetic model pointed out heterogeneities in the waste deposits located on the top layers, where highly conductive signal is associated with different waste composition, organic waste, or a higher moisture content. The electrical conductivity decreases with depth, probably due to the loss of water caused by compaction induced by the weight of the overburden mass. No significant conductivity anomalies have been observed in the expected transition depth between the upper waste deposits and the lower bedrock. This evidence leads to exclude ruptures of the HDPE liner placed at the bottom of the landfill because a leachate migration below the landfill bottom would cause large electrical conductivity anomalies in the bedrock. The Asiago landfill case shows how a careful planning of geophysical measurements is essential in order to produce the best strategy for achieving the set goal in such complex sites, and in particular in order to confirm/dismiss an hypothesis concerning the landfill and its subsurface, in this case the presence of a large karstic cavity. In addition, the survey showed also some unexpected results, i.e., the presence of a low-resistivity anomaly that, in the context of landfills, is a general warning sign, as leachate may produce low conductivity anomalies in the surrounding of a landfill, in presence of a leakage. In this case, this anomaly seen by geophysics has driven a direct investigation that ensured that the anomaly itself has no relationship with the landfill. Overall this is an example of a virtuous circle between noninvasive and invasive investigations, as it should be. The results encourage the use of geophysical data, both for static characterization and dynamic monitoring as good practices for checking the health condition of a landfill facility in order to minimize public health risks, ensure environmental protection and avoid expensive reclamation activities.

Integrating the geophysical information with routine environmental matrices monitoring as defined by the community regulations can represent an optimal strategy for the development of a site conceptual model essential in the management of waste landfill both in operational and postoperational stages.

## Acknowledgements

The authors wish to thank the managers of the three landfill sites (Monteco srl, manager of the Ugento landfill, Progetto Ambiente Bacino Lecce DUE s.r.l, manager of the Corigliano d'Otranto landfill, and the Alto Vicentino Ambiente manager of the Asiago landfill) for making the facility available to carry out the geophysical measurements. Moreover, all authors wish to thank the Apulian Regional Agency for the Protection of the Environment (ARPA PUGLIA) for funding the research at the Ugento landfill, the Municipality of Asiago (VI) for funding the research at the Asiago landfill, for granting permission for the publication of the results and funding direct drilling investigation, the Progetto Ambiente Bacino Lecce DUE s.r.l for funding the research activities at the Corigliano d'Otranto landfill. In addition, the authors wish to thank Celeste Antonietta Turturro, Jacopo Boaga, and Benjamin Mary for supporting the field experimental activity.

## Conflict of interest

The authors declare no conflict of interest.

## Author details

Lorenzo De Carlo[1*], Giorgio Cassiani[2,3], Rita Deiana[3,4], Gian Piero Deidda[5] and Maria Clementina Caputo[1]

1 Water Research Institute (IRSA), National Research Council (CNR), Bari, Italy

2 Department of Geosciences, University of Padova, Padova, Italy

3 Interdepartmental Center for Cultural Heritage—CIBA—University of Padova, Padova, Italy

4 Department of Cultural Heritage, University of Padova, Padova, Italy

5 Department of Civil and Environmental Engineering and Architecture, University of Cagliari, Italy

*Address all correspondence to: lorenzo.decarlo@cnr.it

## References

[1] Soupios P, Papadopoulos I, Kouli M, Georgaki I, Vallianatos F, Kokkinou E. Investigation of waste disposal areas using electrical methods: A case study from Chania, Crete, Greece. Environmental Geology. 2007;**51**:1249-1261. DOI: 10.1007/s00254-006-0418-7

[2] Chambers JE, Kuras O, Meldrum PI, Ogilvy RD, Hollands J. Electrical resistivity tomography applied to geologic, hydrogeologic, and engineering investigations at a former waste-disposal site. Geophysics. 2006;**71**:B231-B239. DOI: 10.1190/1.2360184

[3] Gazoty A, Fiandaca G, Pedersen J, Auken E, Christiansen AV. Mapping of landfills using time-domain spectral induced polarization data: The Eskelund case study. Near Surface Geophysics. 2012;**10**:575-586. DOI: 10.3997/1873-0604.2012046

[4] Meju MA. Geoelectrical investigation of old/abandoned, covered landfill sites in urban areas: Model development with a genetic diagnosis approach. Journal of Applied Geophysics. 2000;**44**:115-150. DOI: 10.1016/S0926-9851(00)00011-2

[5] Frid V, Doudkinski D, Liskevich G, Shafran E, Averbakh A, Korostishevsky N, et al. Geophysical–geochemical investigation of fire-prone landfills. Environmental Earth Sciences. 2010;**60**:787-798. DOI: 10.1007/s12665-009-0216-0

[6] Frid V, Liskevich G, Doudkinski D, Korostishevsky N. Evaluation of landfill disposal boundary by means of electrical resistivity imaging. Environmental Geology. 2008;**53**:1503-1508. DOI: 10.1007/s00254-007-0761-3

[7] Casado I, Mahjoub H, Lovera R, Fernandez J, Casas A. Use of electrical tomography methods to determinate the extension and main migration routes of uncontrolled landfill leachates in fractured areas. Science of the Total Environment. 2015;**506**:546-553. DOI: 10.1016/j. scitotenv.2014.11.068

[8] Deidda GP, Himi M, Barone I, Cassiani G, Casas PA. Frequency-domain electromagnetic mapping of an abandoned waste disposal site: A case in Sardinia (Italy). Remote Sensing. 2022; **14**:878. DOI: 10.3390/rs14040878

[9] De Iaco R, Green AG, Maurer HR, Horstmeyer H. A combined seismic reflection and refraction study of a landfill and its host sediments. Journal of Applied Geophysics. 2003;**52**:139-156. DOI: 10.1016/S0926-9851(02)00255-0

[10] Lanz E, Maurer H, Green A.G. Refraction tomography over a buried waste disposal site. Geophysics 1998:63: 1414–1433. DOI: 10.1190/1.1444443

[11] Abreu AES, Gandolfo OCB, Vilar OM. Characterizing a Brazilian sanitary landfill using geophysical seismic techniques. Waste Management. 2016;**53**:116-127. DOI: 10.1016/j.wasman.2016.03.048

[12] Di Maio R, Fais S, Ligas P, Piegari E, Raga R, Cossu R. 3D geophysical imaging for site-specific characterization plan of an old landfill. Waste Management. 2018;**76**:629-642. DOI: 10.1016/j. wasman.2018.03.004

[13] Cardarelli E, Bernabini M. Two case studies of the determination of parameters of urban waste dumps. Journal of Applied Geophysics. 1997;**36**:167-174. DOI: 10.1016/S0926-9851(96)00056-0

[14] Dumont G, Robert T, Marck N, Nguyen F. Assessment of multiple geophysical techniques for the

characterization of municipal waste deposit sites. Journal of Applied Geophysics. 2017;**145**:74-83. DOI: 10.1016/j.jappgeo.2017.07.013

[15] Kondracka M, Stan-Kłeczek I, Sitek S, Ignatiuk D. Evaluation of geophysical methods for characterizing industrial and municipal waste dumps. Waste Management. 2021;**125**:27-39. DOI: 10.1016/j.wasman.2021.02.015

[16] Leroux V, Dahlin T, Svensson M. Dense resistivity and induced polarization profiling for a landfill restoration project at Harlov, Southern Sweden. Waste Management and Research. 2007;**25**:49-60. DOI: 10.1177/0734242X07073668

[17] Carpenter PJ, Calkin SF, Kaufmann RS. Assessing a fractured landfill cover using electrical resistivity and seismic refraction techniques. Geophysics. 1991;**56**:1896-1904. DOI: 10.1190/1.1443001

[18] White CC, Barker RD. Electrical leak detection system for landfill liners: A case history. Ground Water Monitoring and Remediation. 1997;**17**:153-159. DOI: 10.1111/j.1745-6592.1997.tb00590.x

[19] Binley A, Daily W, Ramirez A. Detecting leaks from waste storage ponds using electrical tomographic methods. In: Proceedings of the 1st World Congress on Industrial Process Tomography; 14–17 April 1999. Greater Manchester: Buxton; 1999, 1999. pp. 6-13

[20] Binley A, Daily W, Ramirez A. Detecting leaks from environmental barriers using electrical current imaging. Journal of Environmental and Engineering Geophysics. 1997;**2**:11-19. DOI: 10.4133/JEEG2.1.11

[21] Colucci P, Darilek GT, Laine DL, Binley A. Locating landfill leaks covered with waste. In: Proceedings of the Sardinia 99, Seventh International Waste Management and Landfill Symposium; 4–8 October 1999. Cagliari, Italy: CISA, Environmental Sanitary Engineering Centre; 1999. pp. 137-140

[22] Frangos W. Electrical detection of leaks in lined waste disposal ponds. Geophysics. 1997;**62**:1737-1744. DOI: 10.1190/1.1444274

[23] Laine DL, Binley AM, Darilek GT. How to Locate Liner Leaks Under Waste, Geotechnical Fabrics Report, 1997. Available from: https://trid.trb.org/view/576599 [Accessed: February 10, 2023]

[24] Laine DL, Darilek GT, Binley AM. Locating Geomembrane liner leaks under waste in a landfill. In: Proceedings of the International Geosynthetics '97 Conference; 11–13 March 1997. Long Beach, USA; 1997. pp. p407-p411

[25] Parra JO. Electrical response of a leak in a geomembrane liner. Geophysics. 1988; **53**:1445-1452. DOI: 10.1190/1.1442424

[26] Binley A, Daily W. The performance of electrical methods for assessing the integrity of geomembrane liners in landfill caps and waste storage ponds. Journal of Environmental and Engineering Geophysics. 2003;**8**:227. DOI: 10.4133/JEEG8.4.227

[27] De Carlo L, Perri MT, Caputo MC, Deiana R, Vurro M, Cassiani G. Characterization of a dismissed landfill via electrical resistivity tomography and mise-à-la-masse method. Journal of Applied Geophysics. 2013;**98**:1-10. DOI: 10.1016/j.jappgeo.2013.07.010

[28] Cassiani G, Fusi N, Susanni D, Deiana R. Vertical radar profiles for the assessment of landfill capping effectiveness. Near Surface Geophysics.

2008;**6**:133-142. DOI: 10.3997/1873-0604.2008010

[29] Deidda GP, De Carlo L, Caputo MC, Cassiani G. Frequency domain electromagnetic induction imaging: An effective method to see inside a capped landfill. Waste Management. 2022;**144**: 29-40. DOI: 10.1016/j.wasman.2022.03.007

[30] Zume JT, Tarhule A, Christenson S. Subsurface imaging of an abandoned solidwaste landfill site in Norman, Oklahoma. Ground Water Monitoring and Remediation. 2006;**26**:62-69. DOI: 10.1111/j.1745-6592.2006.00066.x

[31] Radulescu M, Valerian C, Yang J. Time-lapse electrical resistivity anomalies due to contaminant transport around landfills. Annals of Geophysics. 2007;**50**:453-468. DOI: 10.4401/ag-3075

[32] Ogilvy R, Meldrum P, Chambers J, Williams G. The use of 3D electrical resistivity tomography to characterise waste and leachate distributions within a closed landfill Thriplow, UK. Journal of Environmental and Engineering Geophysics. 2002;**7**:11-18. DOI: 10.4133/JEEG7.1.11

[33] Lopes DD, Silva SMCP, Fernandes F, Teixeira RS, Celligoi A, Dall'Antônia LH. Geophysical technique and groundwater monitoring to detect leachate contamination in the surrounding area of a landfill—Londrina (PR—Brazil). Journal of Environmental Management. 2012;**113**:481-487. DOI: 10.1016/j.jenvman.2012.05.028

[34] Nobes DC, Armstrong MJ, Close ME. Delineation of a landfill leachate plume and flow channels in coastal sands near Christchurch, New Zealand, using a shallow electromagnetic survey method. Hydrogeology Journal. 2000;**8**:328-336. DOI: 10.1007/s100400050018

[35] Triantafilis J, Roe JAE, Monteiro Santos FA. Detecting a leachate plume in an Aeolian sand landscape using a DUALEM-421 induction probe to measure electrical conductivity followed by inversion modeling. Soil Use and Management. 2011;**27**:357-366. DOI: 10.1111/j.1475-2743.2011.00352.x

[36] Bavusi M, Rizzo E, Lapenna V. Electromagnetic methods to characterize the Savoia di Lucania waste dump (southern Italy). Environmental Geology. 2006;**51**:301-308. DOI: 10.1007/s00254-006-0327-9

[37] Monteiro Santos FA, Mateus A, Figueiras J, Gonçalves MA. Mapping groundwater contamination around a landfill facility using the VLF-EM method — A case study. Journal of Applied Geophysics. 2006;**60**:115-125. DOI: 10.1016/j.jappgeo.2006.01.002

[38] Porsani JL, Filho WM, Elis VR, Shimeles F, Dourado JC, Moura HP. The use of GPR and VES in delineating a contamination plume in a landfill site: A case study in SE Brazil. Journal of Applied Geophysics. 2004;**55**:199-209. DOI: 10.1016/j.jappgeo.2003.11.001

[39] Faraco Gallas JD, Taioli F, Malagutti FW. Induced polarization, resistivity, and self-potential: A case history of contamination evaluation due to landfill leakage. Environmental Earth Sciences. 2011;**63**:251-261. DOI: 10.1007/s12665-010-0696-y

[40] Soupios P, Papadopoulos N, Papadopoulos I, Kouli M, Vallianatos F, Sarris A, et al. Application of integrated methods in mapping waste disposal areas. Environmental Geology. 2007;**53**:661-675. DOI: 10.1007/s00254-007-0681-2

[41] Audebert M, Clément R, Moreau S, Duquennoi C, Loisel S, Touze-Foltz N. Understanding leachate flow in

municipal solid waste landfills by combining time-lapse ERT and subsurface flow modelling—Part I: Analysis of infiltration shape on two different waste deposit cells. Waste Management. 2016;**55**:165-175. DOI: 10.1016/j.wasman.2016.04.006

[42] Audebert M, Oxarango L, Duquennoi C, Touze-Foltz N, Forquet N, Clément R. Understanding leachate flow in municipal solid waste landfills by combining time-lapse ERT and subsurface flow modelling—Part II: Constraint methodology of hydrodynamic models. Waste Management. 2016;**55**:176-190. DOI: 10.1016/j.wasman.2016.04.005

[43] Acworth RI, Jorstad LB. Integration of multi-channel piezometry and electrical tomography to better define chemical heterogeneity in a landfill leachate plume within a sand aquifer. Journal of Contaminant Hydrology. 2006;**83**:200-220. DOI: 10.1016/j.jconhyd.2005.11.007

[44] Aristodemou E, Thomas-Betts A. DC resistivity and induced polarisation investigations at a waste disposal site and its environments. Journal of Applied Geophysics. 2000;**44**:275-302. DOI: 10.1016/S0926-9851(99)00022-1

[45] Al-Tarazi E, Abu Rajab J, Al-Naqa A, El-Waheidi M. Detecting leachate plumes and groundwater pollution at Ruseifa municipal landfill utilizing VLF-EM method. Journal of Applied Geophysics. 2008;**65**:121-131. DOI: 10.1016/j.jappgeo.2008.06.005

[46] Hu X, Han Y, Wang Y, Zhang X, Du L. Experiment on monitoring leakage landfill leachate by parallel potentiometric monitoring method. Scientific Reports. 2022;**12**:20496. DOI: 10.1038/s41598-022-24352-w

[47] Grellier S, Guérin R, Robain H, Bobachev A, Vermeersch F, Tabbagh A. Monitoring of leachate recirculation in a bioreactor landfill by 2-D electrical resistivity imaging. Journal of Environmental and Engineering Geophysics. 2008;**13**:351-359. DOI: 10.2113/JEEG13.4.351

[48] Konstantaki LA, Ghose R, Draganov D, Diaferia G, Heimovaara T. Characterization of a heterogeneous landfill using seismic and electrical resistivity data. Geophysics. 2015;**80**: EN13-EN25. DOI: 10.1190/GEO2014-0263.1

[49] Hermozilha H, Grangeia C, Matias MS. An integrated 3D constant offset GPR and resistivity survey on a sealed landfill — Ilhavo, NW Portugal. Journal of Applied Geophysics. 2010;**70**: 58-71. DOI: 10.1016/j.jappgeo.2009.11.004

[50] Mota R, Monteiro Santos FA, Mateus A, Marques FO, Gonçalves MA, Figueiras J, et al. Granite fracturing and incipient pollution beneath a recent landfill facility as detected by geoelectrical surveys. Journal of Applied Geophysics. 2004;**57**:11-22. DOI: 10.1016/j.jappgeo.2004.08.007

[51] Olofsson B, Jernberg H, Rosenqvist A. Tracing leachates at waste sites using geophysical and geochemical modelling. Environmental Geology. 2006;**49**:720-732. DOI: 10.1007/s00254-005-0117-9

[52] Osiensky JL. Ground water modeling of mise-à-la-masse delineation of contaminated ground water plumes. Journal of Hydrology. 1997;**197**:146-165. DOI: 10.1016/S0022-1694(96)03279-9

[53] Manzur SR, Hossain MS, Kemler V, Khan MS. Monitoring extent of moisture variations due to leachate recirculation

in an ELR/bioreactor landfill using resistivity imaging. Waste Management. 2016;**55**:38-48. DOI: 10.1016/j.wasman.2016.02.035

[54] Clément R, Descloitres M, Günther T, Oxarango L, Morra C, Laurent JP, et al. Improvement of electrical resistivity tomography for leachate injection monitoring. Waste Management. 2010;**30**:452-464. DOI: 10.1016/j.wasman.2009.10.002

[55] Clément R, Oxarango L, Descloitres M. Contribution of 3-D time-lapse ERT to the study of leachate recirculation in a landfill. Waste Management. 2011;**31**:457-467. DOI: 10.1016/j.wasman.2010.09.005

[56] Guérin R, Munoz ML, Aran C, Laperrelle C, Hidra M, Drouart E, et al. Leachate recirculation: Moisture content assessment by means of a geophysical technique. Waste Management. 2004; **24**:785-794. DOI: 10.1016/j.wasman.2004.03.010

[57] Audebert M, Clément R, Grossin-Debattista J, Günther T, Touze-Foltz N, Moreau S. Influence of the geomembrane on time-lapse ERT measurements for leachate injection monitoring. Waste Management. 2014; **34**:780-790. DOI: 10.1016/j.wasman.2014.01.011

[58] Degueurce A, Clément R, Moreau S, Peu P. On the value of electrical resistivity tomography for monitoring leachate injection in solid state anaerobic digestion plants at farm scale. Waste Management. 2016;**56**:125-136. DOI: 10.1016/j.wasman.2016.06.028

[59] Jansen J, Haddad B, Fassbender W, Jurcek P. Frequency domain electromagnetic induction sounding surveys for landfill site characterization studies. Groundwater Monitoring and Remediation. 1992;**12**:103-109. DOI: 10.1111/j.1745-6592.1992.tb00068.x

[60] Yannah M, Martens K, Van Camp M, Walraevens K. Geophysical exploration of an old dumpsite in the perspective of enhanced landfill mining in Kermt area, Belgium. Bulletin of Engineering Geology and the Environment. 2019;**78**:55-67. DOI: 10.1007/s10064-017-1169-2

[61] Yochim A, Zytner RG, McBean EA, Endres AL. Estimating water content in an active landfill with the aid of GPR. Waste Management. 2013;**33**:2015-2028. DOI: 10.1016/j.wasman.2013.05.020

[62] Kibria G, Hossain MS. Investigation of degree of saturation in landfill liners using electrical resistivity imaging. Waste Management. 2015;**29**:197-204. DOI: 10.1016/j.wasman.2015.02.015

[63] Steiner M, Katona T, Fellner J, Flores OA. Quantitative water content estimation in landfills through joint inversion of seismic refraction and electrical resistivity data considering surface conduction. Waste Management. 2022;**149**:21-32. DOI: 10.1016/j.wasman.2022.05.020

[64] Dumont G, Pilawski T, Dzaomuho-Lenieregue P, Hiligsmann S, Delvigne F, Thonart P, et al. Gravimetric water distribution assessment from geoelectrical methods (ERT and EMI) in municipal solid waste landfill. Waste Management. 2016;**55**:129-140. DOI: 10.1016/j. wasman.2016.02.013

[65] Imhoff PT, Reinhart DR, Englund M, Guerin R, Gawande N, Han B, et al. Review of state of the art methods for measuring water in landfills. Waste Management. 2007;**27**:729-745. DOI: 10.1016/j.wasman.2006.03.024

[66] Flores-Orozco A, Gallistl J, Steiner M, Brandstätter C, Fellner J.

Mapping biogeochemically active zones in landfills with induced polarization imaging: The Heferlbach landfill. Waste Management. 2020;**107**:121-132. DOI: 10.1016/j.wasman.2020.04.001

[67] Georgaki I, Soupios P, Sakkas N, Ververidis F, Trantas E, Vallianatos F, et al. Evaluating the use of electrical resistivity imaging technique for improving CH 4 and CO 2 emission rate estimations in landfills. Science of the Total Environment. 2008;**389**:522-531. DOI: 10.1016/j.scitotenv.2007.08.033

[68] Johansson S, Rosqvist H, Svensson M, Dahlin T, Leroux V. An alternative methodology for the analysis of electrical resistivity data from a soil gas study. Geophysical Journal International. 2011;**186**:632-640. DOI: 10.1111/j.1365-246X.2011.05080

[69] Noguera JF, Rivero L, Font X, Navarro A, Martínez F. Simultaneous use of geochemical and geophysical methods to characterise abandoned landfills. Environmental Geology. 2002; **41**:898-905. DOI: 10.1007/s00254-001-0467-x

[70] Senos Matias M, Marques da Silva M, Ferreira P, Ramalho E. A geophysical and hydrogeological study of aquifers contamination by a landfill. Journal of Applied Geophysics. 1994;**32**:155-162. DOI: 10.1016/0926-9851(94)90017-5

[71] Binley A, Kemna A. DC resistivity and induced polarization methods. In: Rubin Y, Hubbard SS, editors. Hydrogeophysics. 1st ed. Dordrecht: Springer; 2005. pp. 129-156

[72] Schlumberger C. Ètude sur la prospection electrique du sous-sol. 1st ed. Paris: Gauthier-Villars; 1920. pp. 1-94

[73] Parasnis DS. Three-dimensional electric mise-a-la-masse survey of an irregular lead–zinc–copper deposit in Central Sweden. Geophysical Prospecting. 1967;**15**:407-437. DOI: 10.1111/j.1365-2478.1967.tb01796.x

[74] Ketola M. Some points of view concerning mise-a-la-masse measurements. Geoexploration. 1972;**10**: 1-21. DOI: 10.1016/0016-7142(72) 90010-5

[75] Bowker A. Size determination of slab-like ore bodies an interpretation scheme for single hole mise-a-la-masse anomalies. Geoexploration. 1987;**24**: 207-218. DOI: 10.1016/0016-7142(87) 90028-7

[76] McNeill JD. Electromagnetic terrain conductivity measurements at low induction numbers. Technical Note TN-6. 1980. Available from: http://www.geonics.com/pdfs/technicalnotes/tn6.pdf [Accessed: 10 January 2020]

[77] Binley A, Ramirez A, Daily W. Regularised image reconstruction of noisy electrical resistance tomography data. In: Proceedings of the 4th Workshop of the European Concerted Action on Process Tomography. Bergen: UMIST, University of Manchester Institute of Science and Technology; 1995. pp. 401-410

[78] Deidda GP, Díaz de Alba P, Fenu C, Lovicu G, Rodriguez G. FDEMtools: A MATLAB package for FDEM data inversion. Numerical Algorithms. 2020; **84**:1313-1327. DOI: 10.1007/s11075-019-00843-2

Chapter 8

# Waste in the Treatment of Textile Wastewater by Pressure-Driven Membrane Processes

*Iva Ćurić, Luka Brezinščak and Davor Dolar*


## Abstract

Due to strong globalization and industrialization, water has become a scarce resource. One industry that uses a lot of water and generates a large amount of wastewater is the textile industry. According to the Best Available Techniques reference document, pressure-driven membrane processes have been declared the best methods for the treatment and reuse of textile wastewater. Such processes generate a certain amount of solid waste in addition to excellent permeate quality. This book chapter provides a critical overview of pressure-driven membrane processes (microfiltration, ultrafiltration, nanofiltration, reverse osmosis) and membrane bioreactor (MBR) for the treatment of textile wastewater. Finally, this chapter covers the treatment and disposal of retentate and MBR sludge.


**Keywords:** pressure-driven membrane processes, solid waste, retentate, sludge, disposal, membrane bioreactor, microfiltration, ultrafiltration, nanofiltration, reverse osmosis, textile wastewater

## 1. Introduction

The increase in population and the growing demand for industry and its products have led to significant impacts on the environment. Textiles are part of people's daily needs, and the "buy, wear, and throw away" model is leading to an increasing demand for textile production [1, 2]. Among the manufacturing industries, the textile industry is considered the most complicated industry because it consists of a large number of production stages [3]. Each of these stages generates a larger amount of dry or wet waste. Therefore, production in the textile industry is divided into dry and wet processes. Dry processes produce unfinished textiles, while wet processes produce wastewater, which is generated by the combination of fresh water, which serves as a medium, and certain chemicals (auxiliaries, alkalis, acids, salts) [4]. The production of 1 kg of textiles consumes about 150 L of fresh water [5]. However, the amount of water depends on the type of work process, the machines used in production, and the type of material [6]. The wastewater that remains at the end of the process contains large concentrations of pollutants that cause significant concentrations of physico-chemical (chemical oxygen demand, biological oxygen demand, total organic carbon, etc.), ecotoxicological (acute and chronic toxicity), and inorganic parameters (presence of metals) [7]. All this can have negative effects on the environment and life in the environment. For example, textile wastewater can disrupt photosynthesis and have carcinogenic and mutagenic effects on aquatic life [8]. Therefore, it is necessary to treat textile wastewater before it is discharged into natural recipients. Over the years, conventional textile wastewater treatment processes (adsorption, coagulation, etc.) have been developed, but they have problems with sludge formation, high investment costs, frequent use of chemicals, and many other problems [7, 9]. To minimize these problems, pressure-driven membrane processes (microfiltration (MF), ultrafiltration (UF), nanofiltration (NF), and reverse osmosis (RO)) can be used. These processes have been shown to be a high-quality option for obtaining a high-quality effluent that meets state regulations (if any) for discharge. They have also shown that the treated wastewater can be reused in production processes, reducing the need for fresh water [9]. Each technology has its advantages and disadvantages, for example, pressure-driven membrane processes "struggle" with high energy consumption, greenhouse gas emissions, and the generation of waste, the retentate [10]. The retentate is one of the streams in cross-flow separation that contains higher concentrations of pollutants than the feed stream. It is necessary to treat the retentate before disposal because it has disastrous consequences for the environment [11, 12]. In membrane bioreactors (MBRs), a certain amount of sludge is produced due to the growth and decay of the bacteria contained in them [13]. In this case, treatment and disposal is required to avoid negative impacts on the environment.

The authors have not found any study on the characterization of the retentate and MBR sludge after the treatment of textile wastewater or on the possibility of its disposal. Therefore, this study presents the possibilities of treatment and disposal of retentate and MBR sludge from desalination plants and municipal wastewater treatment plants. This opens space for new research possibilities on the topic of retentate and MBR sludge after treatment of textile wastewater.

## 2. Textile industry

Industrialization contributes to economic growth by generating income and providing jobs for a growing population. It is the main driver of poverty reduction and shared prosperity. Therefore, industry plays a key role in global economic growth in both developed and developing countries [1, 14]. One of the largest industries contributing to economic growth is the textile industry, which is constantly expanding. With about 120 million employees and a global market share of about $2000 billion, the textile industry is an important contributor to the global economy [15]. Apart from its importance to the economy, the textile industry is also important to every human being as it contributes to the fulfillment of basic necessities of life, namely clothing [2]. To obtain a single textile product, it is necessary to pass through some of the largest and most complicated industrial chains among other various industrial products, which is characteristic of the textile industry [3]. The complex characterization of textile products is due to the wide range of chemicals, process sequences, various machines, and finishes [6].

The production chain begins with the drying process, i.e., the transformation of natural (cotton, linen, etc.) or artificial (polyester, polyamide, etc.) raw materials into yarn by spinning processes. The yarn is then used to knit or weave to produce knitted or woven fabrics. Depending on customer requirements, wet processes follow, consisting mainly of pretreatment, mercerizing, washing, bleaching, dyeing, finishing, and printing. At the end are squeezing, drying, calendering, tailoring, and sewing. Dry processing produces solid waste, while wet processing steps mainly produce liquid waste [16].

### 2.1 Textile wastewater

To improve the properties of textiles, the use of wet processes is necessary [17]. These processes use water, chemicals, dyes, and auxiliaries. It is estimated that the use of auxiliaries accounts for 60–70% of dyes, while the consumption of dyes per kg of product is 50–90 g [18]. Water is the most important liquid substance for performing wet processes and has three functions: as a solvent for dyes and chemicals, as a medium for transferring dyes and chemicals to textiles, and as a medium for washing and rinsing [19]. Fresh water is very commonly used, and the average consumption is 150 L per 1 kg of product but can be as high as 933 L per 1 kg of textile product [5]. At the end of each wet process, a large amount of wastewater is generated, consisting of unfixed dyes, auxiliaries, and other chemicals. Due to its composition, textile wastewater contains a high color, biological oxygen demand (BOD), chemical oxygen demand (COD), pH, total suspended solids (TSS), total dissolved solids (TDS), etc. [20]. The World Bank estimates that the textile industry discharges 1–10 million liters of wastewater per day, and about 17–20% of it is discharged into the environment without prior treatment [1].

### 2.2 Textile wastewater treatment

Wastewater from the textile industry can actually disturb the natural balance if it is discharged directly into the environment without prior treatment. Ministry of Environmental Protection and Energy in Croatia has set the minimum permissible concentrations for wastewater emissions into natural aquifers and public drainage systems, depending on local conditions and environmental protection requirements. Unfortunately, not every state has legally permissible concentrations for discharges, which is very problematic [21]. Treatment of textile wastewater can be challenging because the toxic pollutants mentioned above are very difficult to degrade in wastewater. In recent decades, many techniques have been developed to find an economical and efficient way to treat textile wastewater, including physical, chemical, biological, combined treatment processes, and other technologies [7]. Equalization and homogenization, flotation, adsorption, ion exchange, and others have been applied as physical processes, while chemical coagulation, electrocoagulation, and oxidation treatment have been used as chemical processes. Biological methods consist of aerobic and anaerobic oxidation. These methods have disadvantages, such as the production of a large amount of concentrated sludge, some are very expensive, require the addition of chemicals, have large footprint, and are not suitable for every textile wastewater [9]. Some of the methods cannot meet the legal limits, which are increased every year.

## 3. Pressure-driven membrane processes

In terms of sustainability and applicability, membrane technology shows great potential when all textile wastewater treatments are considered. Membrane technology fits into a concept of clean production based on the concept of Best Available Techniques (BAT), which must satisfy legal, environmental, and technical requirements [22]. Membrane technology fulfills one of two main strategies: efficient technology for end-of-pipe treatment [23]. This advanced treatment technology enables the treatment of wastewater in a small space and is very easy to apply, which enables a high level of wastewater treatment. In addition, it enables a successful transition from laboratory to pilot or full scale [9].

This type of technology is not a new invention, and the diversity of textile wastewater also presents a challenge for this technology. However, unlike conventional textile wastewater treatment processes, improvements can be made by modifying the membrane modules and elements. Singh and Hankins stated that membrane technology can bridge the gap between cost-effectiveness and sustainability with little or no use of chemicals, which is evidence of its environmental friendliness [24]. Also worth mentioning is the possibility of hybrid treatment, i.e., combining different membrane processes or conventional treatments with membrane processes [9].

Among other membrane technologies, pressure-driven membrane processes are most commonly used in wastewater treatment from pretreatment to post-treatment. The treatment principle is based on the application of hydraulic pressure to water, which thus passes through a porous membrane that retains pollutants (**Figure 1**), i.e., the feed stream is divided into two streams. Permeate is the stream that passes through the membrane ("clean water"), and retentate is the stream that is retained by the membrane. There are four main types of these processes: MF, UF, NF, and RO. The main differences between these processes are shown in **Figure 2** and **Table 1** [25–27].

### 3.1 Membrane bioreactor

Pressure-driven membrane processes not only enable hybrid performance but have also proven effective in combination with conventional processes, i.e., biological treatment. In the late 1960s, the first MBR was commercially developed by Dorr-Oliver. Both capital and operating costs were initially very high, but in the last 15 years, costs have been significantly reduced [12]. The unique feature of this

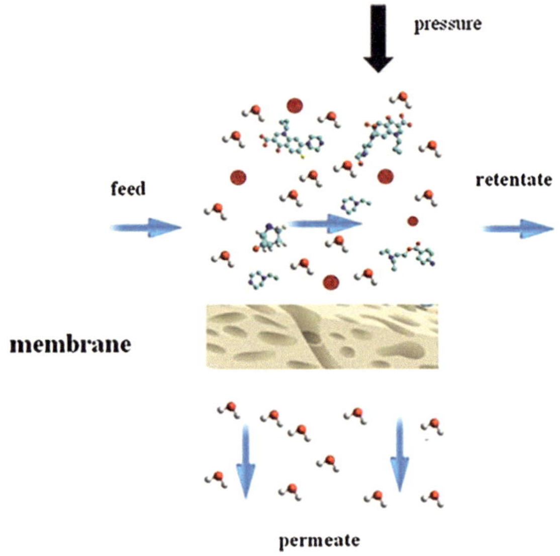


**Figure 1.**
*Pressure-driven membrane processes.*

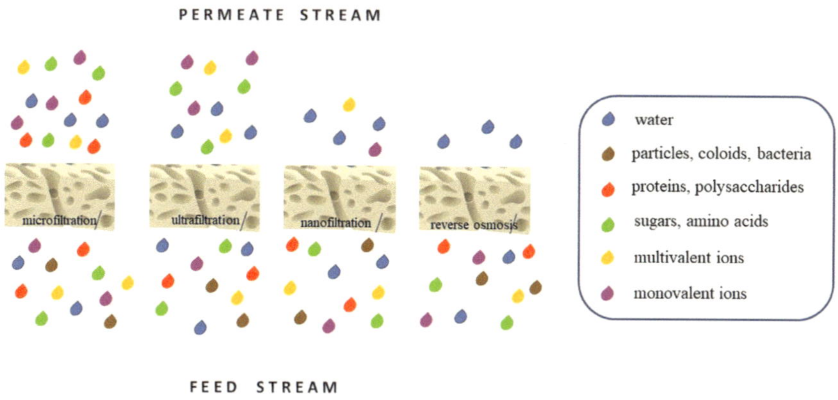


**Figure 2.**
*Pressure-driven membrane processes (MF, UF, NF, RO).*

| Process | Membrane type | Pore radius/μm | MWCO/kDa | Pressure/MPa |
|---|---|---|---|---|
| MF | Symmetric porous | 0.1–10 | 100–500 | 0.05–0.2 |
| UF | Asymmetric porous | 2–10 | 20–150 | 0.1–1 |
| NF | Asymmetric tight porous | 0.002–0.05 | 2–20 | 0.5–4 |
| RO | Asymmetric skin-type solution-diffusion | 0.0001–0.001 | 0.2–2 | 1–10 |

**Table 1.**
*Main pressure-driven membrane processes data.*

technology is that it involves two separate processes. The first is biological treatment, and the second is membrane separation. In contrast to conventional biological treatment, MBR is characterized by consistent water quality, small footprint, reduced sludge production, and high pollutant removal rates. The Institute of Textile Research and Industrial Cooperation in Terrassa has conducted comparative studies on the treatment of textile wastewater with MBR and conventional biological systems [28]. The results are presented in **Table 2**. They showed that MBE has better efficiency in terms of COD and color removal from textile wastewater with lower sludge production. MBR can be operated in aerobic (AeMBR) or anaerobic (AnMBR) mode. In general, AnMBR requires longer sludge retention time, which may lead to greater fouling problems than AeMBR [29]. Friha et al. removed COD, color, SS, and $BOD_5$ at 96–100% with AeMBR [30]. Yurtsever et al. achieved complete color removal from textile wastewater with AnMBR and 30–50% with AeMBR [31].

| Type of the process | COD removal (%) | Color removal (%) |
|---|---|---|
| Conventional biological process | 70 | < 92 |
| MBR | 81 | 92 |

**Table 2.**
*Comparative study of textile wastewater treatment by conventional biological process and MBR.*

## 4. Challenges in purification of textile wastewater

Pressure-driven membrane processes have proven to be satisfactory replacement for conventional textile wastewater treatment processes. Nevertheless, each technology has its drawbacks. For example, pressure-driven membrane processes have a greenhouse gas problem because they consume large amounts of energy since pressure is the driving force. The biggest problem with this technology is the second stream in the treatment, the retentate [10]. Other names include brine, brine reject, and concentrate, but all refer to the concentrated waste stream [32]. Because the retentate is the stream that is retained by the membrane, it contains much higher concentrations of pollutants than the feed stream (wastewater). In addition to the substances from the feed stream, it may also contain agents for membrane cleaning, maintenance, and optimization of membrane performance [10, 32].

Jones et al. reported that retentate production in 2019 was 141.5 million $m^3$ $day^{-1}$ [33]. Katal et al. noted that approximately 5 to 33% of the total cost of membrane treatment is related to retentate disposal. The exact percentage depends on the characteristics and volume of the retentate and the disposal option [34]. In 2006, Mickley & Associates reported in the second edition of Desalination and Water Purification Research and Development Program Report No. 123, in which they outlined the regulations for retentate disposal in the USA [35]. As far as the authors are aware, no such regulations exist in Europe.

### 4.1 Disposal of the retentate

The management and disposal of retentate from any membrane wastewater treatment plant requires careful consideration to avoid environmental degradation. While there is a considerable amount of knowledge and practical experience related to the discharge of retentate from seawater and brackish water desalination plants, to the authors' knowledge, no study has been conducted on the environmental impacts of discharging retentate from water treatment plants to water bodies or the ocean [36, 37]. When retentate is generated from a wastewater treatment plant, it is undeniable that it contains toxic substances in very high concentrations that affect the environment.

Retentate is usually discharged into sewers, oceans or seas, ponds, and deep wells. Ariono et al. explained the double impact of retentate: physico-chemical and ecological. The physico-chemical properties of the receiving water may change due to high pollutant content or salinity. Salt in water can increase the osmotic pressure, and the disturbance of the equilibrium between organisms and receiving water can damage the cells of organisms [38]. Jenkins et al. found that salinity as low as 2–3 g $L^{-1}$ can have effects on several marine species [39]. In addition to salinity, thermal pollution can also occur from the disposal of retentate. Retentate raises the temperature of about 60% of seawater, resulting in less dissolved oxygen than in cool water. These conditions lead to hypoxia and death of aquatic organisms [40]. Higher alkalinity is another problem in retentate disposal. Ahmet et al. found that the discharge of retentate increases the concentrations of carbonate, sulfate ions, and other elements to twice the normal value [40]. The disposal of retentate on soil can also affect the physico-chemical properties of soil. Anders et al. reported the effects of retentate disposal from desalination plants on soils in western Rio Grande do Norte, Brazil. Analysis of electrical conductivity and sodium saturation in the soil showed that 45% of the samples at the site of retentate discharge were saline and 25% were salic [41].

### 4.2 Solutions for the retentate

As mentioned above, discharge of retentate to surface waters is the most common practice and the cheapest disposal option. Prior to discharge, certain preparatory measures must be taken to avoid endangering surface waters. The authors found only one study dealing with the reuse of retentate from the treatment of textile wastewater, so the disposal of retentate from desalination plants is described below.

A promising solution could be dilution with seawater or municipal wastewater to reduce the salt concentration in the retentate from desalination plants [42]. Malfeito et al. investigated the possibility of diluting the retentate of RO from the desalination plant in Javea, on the Spanish Mediterranean coast, into the beach of the Fontana Channel. The dilution eliminated the problem of anoxia [43]. Shrivastava and Adams diluted retentate from a desalination plant with cooling water from the condenser, treated wastewater, and seawater. Dilution with treated wastewater results in higher effective salinity dilution than predilution with cooling water or seawater [4].

One option for retentate disposal and management is land application. Retentate can be applied when vegetation (hydroponic systems, fields, parks, etc.) requires nutrients. Several issues need to be addressed prior to application. These include the cost of installing an irrigation system, salinity, infiltration rate, and land availability and cost [42].

Jiménez-Arias et al. investigated the possibility of using retentate from desalination plants in the Canary Islands for hydroponic systems for tomato. Dilution of the retentate saved 20% of the cost of the hydroponic solution because the retentate contained nutrient-rich minerals. The tomatoes showed excellent organoleptic quality after cultivation [44].

Saf et al. conducted an integrated study on the agroecological applications of olive mill wastewater from UF and NF. The retentate from both processes was evaluated for phytotoxicity to the two major crops, corn and flax, and to wild mustard germination. Ultrafiltration and NF retentate resulted in a significant increase in corn growth. A study on the weed Sinapis arvensis showed no germination, suggesting a promising potential for the production of bioherbicides from the retentate [45].

Yuzer et al. reused NF retentate as a salt in reactive dyeing at various dilutions after treating textile wastewater for the dyeing process. The spectrophotometric values on textiles showed no deviations [46].

### 4.3 MBR sludge

In Section *3.1.*, we provided information on the significant possibility of using MBR as a promising solution for the treatment of textile wastewater. It is estimated that the average annual production of sludge in Europe, USA, and China is 240 wet tons. Therefore, the treatment and disposal of MBR sludge can be challenging. The U.S. Environmental Protection Agency's Part 503 rule contains comprehensive requirements for the disposal of sludge generated during wastewater treatment [47]. Banti et al. reported a sludge generation of 14 $m^3$ $day^{-1}$ in MBR after treatment of municipal wastewater [48].

Like MBR, conventional activated sludge (CAS) systems are also commonly applied technologies for wastewater treatment [49]. These coupled systems deal with sludge development due to the growth/decay of microorganisms, which is unavoidable in both processes. Apart from the different operating conditions of the two systems (MBR has higher sludge retention time (SRT)), there are also differences in the quality of the resulting sludge at the end of wastewater treatment [13, 49]. Massé et al. and Fenu et al. compared sludge morphology in CAS and MBR systems and showed that the larger size CAS flocs aggregation forms (≈200 μm) consisted of more filamentous bacteria than in MBR, which consisted of smaller flocs (≈50 μm) with more non-flocculating bacteria [50, 51]. Floc aggregation is related to extracellular polymeric substances (EPS), which are the third component between water and the cells most present in the sludge [13]. EPS is the product of bacteria and can account for about 80% of the total mass of the sludge and is responsible for the structural and functional integrity of the flocks [52]. Pontoni et al. found that EPS production in the MBR system is lower than in the CAS system, which is related to the chemical composition of EPS. They reported that the chemical composition of EPS in CAS consisted of proteins (57–61%), followed by carbohydrates (28–29%), humic acids (6–7%), and uronic acids (4–8%). MBR had lower contents of proteins (17–50%), carbohydrates (13–22%), and humic acids (2–4%) at higher uronic acid concentrations (32–60%) [53]. Indeed, the predominant mechanism behind of floc formation depends strongly on the chemical composition of the EPS. Uronic acid in the EPS composition can form hydrogels and interact with the interstitial water in the presence of polyuronate molecules. Polyuronates may hinder flocculation, resulting in a reduction of sludge volume [13].

### *4.3.1 Sludge treatment*

Although the MBR has a lower SRT than the CAS system, the amount of sludge produced does not differ between the two systems when the wastewater supplied and the mass of microorganisms in the reactor are the same. There is a lack of information about MBR sludge treatment and disposal. This is very unusual, as interest in MBR treatments and membrane separation process in general has been increasing over the years. For this reason, it is necessary to know the impact and costs of disposal for the total cost of wastewater treatment [54]. The sludge treatment system must include the following stages before final disposal: 1) thickening, 2) pretreatment, 3) post-treatment stage, and 4) dewatering stage (**Figure 3**). In the thickening stage, the moisture content is removed and the solid content is increased to approximately 2 to 5%. Usually, this can be done by centrifugation, dissolved air flotation, and gravity. This can later help with dewatering [55].

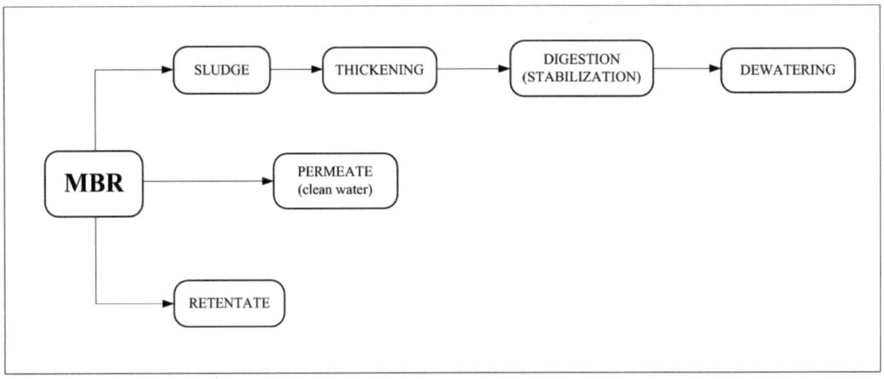


**Figure 3.**
*Sludge treatment steps.*

Dewatering is the reduction of the bound water content, i.e., the reduction of the total volume in order to minimize the costs of sludge treatment and transportation [53]. Dewatering can be accomplished by mechanical dewatering or by thermal dewatering. Mechanical dewatering processes can achieve a maximum solids content of 60%, while a thermal dewatering process has a better solids content of 90% [56]. MBR sludge generally has poorer dewatering properties than CAS sludge, which is due to the previously mentioned different chemical composition of EPS [53].

Before the dewatering process, it is important to reduce the amount of sludge (stabilization). In order to improve the biodegradability of sludge, it is necessary to introduce sludge pretreatment, such as chemical, physical (mechanical, thermal) and biological processes. Pretreatment of sludge reduces particle size and accelerates microbial solubilization, i.e., hydrolysis. Also, during hydrolysis, organic polymers break down into smaller molecules (monomers or dimers) and degrade the EPS. Pretreatments make sludge more susceptible to digestion. Without some kind of pretreatment, the digestion process takes longer and the amount of residual sludge to be disposed of in the end is lower [57]. In addition, it has been found that a higher energy requirement is needed to achieve water removal from sludges with lower water content. Usually, pretreatments affect the metabolism of microorganisms and cell lysis-cryptic growth, leading to the release of intracellular and extracellular substances [13]. Biological methods are more attractive because they are less expensive and more environmentally friendly than chemical methods. To choose the right method, the properties of the raw sludge must be known. Parameters such as pH, organic acid content, and alkalinity can be very important because, for example, the organic acid content can influence the anaerobic digestion process [58].

Aerobic and anaerobic digestion are two well-known biological processes. In aerobic digestion, organic matter is decomposed by aerobic microorganisms using oxygen, while in anaerobic digestion, no oxygen is used, only anaerobic microorganisms. Aerobic digestion is effective in terms of complete pathogen eliminations and short sludge retention, but anaerobic digestion has better cost efficiency and high resource and energy recovery from sludge (conversion of organic matter to biogas such as methane and carbon dioxide) [13]. In addition, anaerobic digestion can reduce the amount to be disposed of while controlling odor emissions [59]. In the European Union, anaerobic digestion is the most common digestion process [60]. Asia et al. subjected textile sludge to anaerobic treatment and obtained a reduction of 61% in total solids, 68% in settleable solids and 51% in volatile solids, 99.99% in total bacteria, BOD and COD by 89% [61]. The authors found no characterization of MBR sludge from textile wastewater treatment and its treatment/disposal. Houghton et al. investigated the effect of ECP on the dewaterability of sludge before and after anaerobic digestion. The ECP content after digestion is lower than before digestion, and the organic composition of ECP changed after digestion with more proteins and carbohydrates. This indicates that the lowering of the ECP content results in a sludge that is easier to dewater [62].

#### *4.3.2 Sludge disposal*

The three most common disposal methods for industrial sludge are landfilling, incineration, and agricultural land application. Landfilling is the most common method in the world because it is simple. Before landfilling, some conditions must be met. In addition to stabilization and dewatering, the solid content of the sludge must be at least 25%. This method cannot be sustainable because environmental standards are constantly increasing and there are problems with the available space for landfilling.

On the other hand, incineration is a thermal waste disposal method that burns the sludge and converts it into gas, ash, and heat that can be used to generate electricity. The incineration method can avoid the problems of heavy metals and pathogens from the sludge that occur during landfilling. However, toxic gasses are released during incineration [55]. Agricultural land application can be beneficial to the soil because sludge is rich in organic matter and contains macro- and micronutrients that can improve plant growth. It also reduces the use of expensive fertilizers. However, sludge can have negative impact on soil quality because it contains pathogenic bacteria and heavy metals that can contaminate human food. In their study, Kakati et al. showed the potential reuse of textile sludge as fertilizer for the growth of green gram. The results showed that sludge at low concentration can be used as fertilizer to promote plant growth, but higher concentration of the sludge may have negative effects on growth due to its toxicity.

Overall, this method is the most economically favored compared to other methods and has fewer global warming problems [63]. These problems can be easily reduced with the above methods (anaerobic digestion, aerobic digestion, and dewatering).

In addition to landfilling, incineration, and disposal on agricultural lands, there are alternative methods of reuse, i.e., solidification (formation of solid material) and use of solid material for a specific type of material. For example, brick and concrete. During solidification, chemical interaction between consolidating additives and impurities does not necessarily occur. Beshah et al. have made bricks from textile sludge. The results show that 10–20% of the bricks meet the standards and were compared to pristine clay bricks [64]. Lissy et al. also used textile sludge to make bricks. Their test results showed that the control bricks and the bricks made from sludge were cast under the same condition, and suggested that the bricks made from sludge should be used for construction purposes [65].

Patel and Pandey used sludge from the CAS after TWW treatment to make blocks as construction material. They investigated the viability of the blocks in terms of solidification time, unconfined compressive quality, and toxicity characteristic. The estimate of the compressive quality of the blocks produced decreased when the percentage of sludge in the cement was increased [66].

Jahagirdar et al. also used textile sludge as an absorbent to remove dyes from wastewater. They burned the sludge at 800°C and studied the ash structure. The ash of the sludge was made permeable, and they found that it could be used as a retention agent for dye-absorbing applications [67].

## 5. Conclusions

Despite the great recognition of membrane pressure-driven processes for wastewater treatment, which are characterized by efficiency and simplicity, they have a problem with the creation of a certain amount of waste after treatment.

This chapter in the book provides information on waste generation during wastewater treatment with membrane pressure-driven processes. Furthermore, the ways of managing this waste are explained. This chapter also provides guidelines on how the generated waste can be recovered.

## Conflict of interest

The authors declare no conflict of interest.

## Notes/thanks/other declarations

I would first like to thank my supervisor/mentor, associate professor Davor Dolar, PhD. I would like to thank him from the bottom of my heart for giving me the guidance and advice I needed to succeed. Working under a great supervisor has been very enjoyable and I have learned and grown a lot.

My second expression of gratitude goes to my colleagues, who have always supported me in my studies and progress. By setting me challenging tasks, they have strengthened me and given me a boost.

Finally, I would like to thank my colleague Luka Brezinščak, mag. ing. agr. who participated with us in this work and future collaboration.

This book chapter is dedicated to my one-year-old son, Fran.

## Author details

Iva Ćurić[1*], Luka Brezinščak[2] and Davor Dolar[1]

1 University of Zagreb Faculty of Chemical Engineering and Technology, Zagreb, Croatia

2 University of Zagreb Faculty of Agriculture, Zagreb, Croatia

*Address all correspondence to: icuric@fkit.unizg.hr

## References


[1] Adane T, Adugna AT, Alemayehu E. Textile industry effluent treatment techniques. Journal of Chemistry. 2021;**2021**:1-14. DOI: 10.1155/2021/5314404

[2] Deng D, Lamssali M, Aryal N, Ofori-Boadu A, Jha MK, Samuel RE. Textiles wastewater treatment technology: A review. Water Environment Research. 2020;**92**(10): 1805-1810. DOI: 10.1002/wer.1437

[3] Paździor K, Bilińska L, Ledakowicz S. A review of the existing and emerging technologies in the combination of AOPs and biological processes in industrial textile wastewater treatment. Chemical Engineering Journal. 2019;**376**:120597. DOI: 10.1016/j.cej.2018.12.057

[4] Shrivastava I, Adams EE. Pre-dilution of desalination reject brine: Impact on outfall dilution in different water depths. Journal of Hydro-Environment Research. 2019;**24**:28-35. DOI: 10.1016/j.jher.2018.09.001

[5] Bilińska L, Gmurek M. Novel trends in AOPs for textile wastewater treatment. Enhanced dye by-products removal by catalytic and synergistic actions. Water Resources and Industry. 2021;**26**:100160. DOI: 10.1016/j.wri.2021.100160

[6] Vhanbatte SB. Background and type of textile materials processed in textile industry. In: Applications of Biotechnology for Sustainable Textile Production. Duxford, United Kingdom: Elsevier; 2022. p. 1-22. DOI: 10.1016/B978-0-323-85651-5.00004-9

[7] Azanaw A, Birlie B, Teshome B, Jemberie M. Textile effluent treatment methods and eco-friendly resolution of textile wastewater. Case Studies in Chemical and Environmental Engineering. 2022;**6**:100230. DOI: 10.1016/j.cscee.2022.100230

[8] Yusuf R, Sonibare J. Characterization of textile industries. Effluents in Kaduna, Nigeria and pollution implications. GlobalNEST. International Journal. 2004;**6**:212-221. DOI: 10.30955/gnj.000284

[9] Keskin B, Ersahin ME, Ozgun H, Koyuncu I. Pilot and full-scale applications of membrane processes for textile wastewater treatment: A critical review. Journal of Water Process Engineering. 2021;**42**:102172. DOI: 10.1016/j.jwpe.2021.102172

[10] Jarma YA, Karaoğlu A, Senan IRA, Baba A, Kabay N. Brine minimization in desalination of the geothermal reinjection fluid by pressure-driven membrane separation processes. Desalination. 2022;**535**:115840. DOI: 10.1016/j.desal.2022.115840

[11] Khan S, Malik A. Toxicity evaluation of textile effluents and role of native soil bacterium in biodegradation of a textile dye. Environmental Science and Pollution Research. 2018;**25**:4446-4458. DOI: 10.1007/s11356-017-0783-7

[12] Judd S, Judd C, editors. Chapter 1 - introduction. In: The MBR Book. Oxford: Elsevier Science; 2006. pp. 1-19. DOI: 10.1016/B978-185617481-7/50003-9

[13] Campo R, Capodici M, Corsino F, Di Bella G, Esposito G, Fabbricino M, et al. 14 - membrane bioreactors sludge: From production to disposal. In: Mannina G, Pandey A, Larroche C, Ng HY, Ngo HH, editors. Current Developments in Biotechnology and Bioengineering. Amsterdam, Netherlands: Elsevier;

2020. pp. 323-351. DOI: 10.1016/B978-0-12-819854-4.00014-9

[14] Upadhyaya S, Kepplinger D. How Industrial Development Matters to the Well-Being of the Population: Some Statistical Evidence. Vienna: UNIDO [Internet]; 2014

[15] Kishor R, Purchase D, Saratale GD, Saratale RG, Ferreira LFR, Bilal M, et al. Ecotoxicological and health concerns of persistent coloring pollutants of textile industry wastewater and treatment approaches for environmental safety. Journal of Environmental Chemical Engineering. 2021;**9**(2):105012. DOI: 10.1016/j.jece.2020.105012

[16] Mani S, Chowdhary P, Bharagava RN. Textile wastewater dyes: Toxicity profile and treatment approaches. Emerging and Eco-Friendly Approaches for Waste Management. 2019;**1**:219-244. DOI: 10.1007/978-981-10-8669-4_11

[17] Kan C-W. Plasma treatments for sustainable textile processing. In: Sustainable Apparel. Cambridge, UK: Elsevier; 2015. pp. 49-118. DOI: 10.1016/B978-1-78242-339-3.00003-0

[18] Sudarshan S, Harikrishnan S, Rathi Bhuvaneswari G, Alamelu V, Aanand S, Rajasekar A, et al. Impact of textile dyes on human health and bioremediation of textile industry effluent using microorganisms: Current status and future prospects. Journal of Applied Microbiology. 2022;**134**(2):1-23. DOI: 10.1093/jambio/lxac064

[19] Senthil Kumar P, Grace PK. 2 - water and textiles. In: Muthu SS, editor. Water in Textiles and Fashion. Duxford, United Kingdom: Woodhead Publishing; 2019. pp. 21-40. DOI: 10.1016/B978-0-08-102633-5.00002-6

[20] Shrivastava R, Singh NK. Assessment of water quality of textile effluent and its treatment by using coagulants and plant material. Materials Today: Proceedings. 2021;**43**:3318-3321. DOI: 10.1016/j.matpr.2021.02.373

[21] Wang Z, Xue M, Huang K, Liu Z. Textile dyeing wastewater treatment. Advances in Treating Textile Effluent. 2011;**5**:91-116. DOI: 10.5772/22670

[22] Ćurić I, Dolar D, Karadakić K. Textile wastewater reusability in knitted fabric washing process using UF membrane technology. Journal of Cleaner Production. 2021;**299**:126899. DOI: 10.1016/j.desal.2009.09.037

[23] Tomei MC, Pascual JS, Angelucci DM. Analysing performance of real textile wastewater bio-decolourization under different reaction environments. Journal of Cleaner Production. 2016;**129**:468-477. DOI: 10.1016/j.jclepro.2016.04.028

[24] Singh R, Hankins N. Emerging Membrane Technology for Sustainable Water Treatment. United Kingdom: Elsevier; 2016

[25] Cirillo AI, Tomaiuolo G, Guido S. Membrane fouling phenomena in microfluidic systems: From technical challenges to scientific opportunities. Micromachines. 2021;**12**(7):820. DOI: 10.3390/mi12070820

[26] Gul A, Hruza J, Yalcinkaya F. Fouling and chemical cleaning of microfiltration membranes: A mini-review. Polymers. 2021;**13**(6):846. DOI: 10.3390/polym13060846

[27] Pal P. Chapter 1 - Introduction to the arsenic contamination problem. In: Pal P, editor. Groundwater Arsenic Remediation. MA, USA: Butterworth-Heinemann; 2015. pp. 1-23. DOI: 10.1016/B978-0-12-801281-9.00001-1

[28] Yang X, Crespi Rosell M, López GV. A review on the present situation of

wastewater treatment in textile industry with membrane bioreactor and moving bed biofilm reactor. Desalination and Water Treatment. 2018;**103**:315-322. DOI: 10.5004/dwt.2018.21962

[29] Jegatheesan V, Pramanik BK, Chen J, Navaratna D, Chang C-Y, Shu L. Treatment of textile wastewater with membrane bioreactor: A critical review. Bioresource Technology. 2016;**204**:202-212. DOI: 10.1016/j.biortech.2016.01.006

[30] Friha I, Bradai M, Johnson D, Hilal N, Loukil S, Ben Amor F, et al. Treatment of textile wastewater by submerged membrane bioreactor: In vitro bioassays for the assessment of stress response elicited by raw and reclaimed wastewater. Journal of Environmental Management. 2015;**160**:184-192. DOI: 10.1016/j.jenvman.2015.06.008

[31] Yurtsever A, Sahinkaya E, Aktaş Ö, Uçar D, Çınar Ö, Wang Z. Performances of anaerobic and aerobic membrane bioreactors for the treatment of synthetic textile wastewater. Bioresource Technology. 2015;**192**:564-573. DOI: 10.1016/j.biortech.2015.06.024

[32] Khan SJ, Murchland D, Rhodes M, Waite TD. Management of concentrated waste streams from high-pressure membrane water treatment systems. Critical Reviews in Environmental Science and Technology. 2009;**39**(5):367-415. DOI: 10.1080/10643380701635904

[33] Jones E, Qadir M, van Vliet MTH, Smakhtin V, Kang S-m. The state of desalination and brine production: A global outlook. Science of The Total Environment. 2019;**657**:1343-1356. DOI: 10.1016/j.scitotenv.2018.12.076

[34] Reza K, Teo Ying S, Iman J, Saeid M-P, Abadi F MHD. An overview on the treatment and Management of the Desalination Brine Solution. In: Davood Abadi F MH, Vahid V, Amir Hooshang T, editors. Desalination. Rijeka: IntechOpen; 2020. p. Ch. 1. DOI: 10.5772/intechopen.92661

[35] Mickley, Associates, Engineering ERTWT, Group R, Research WD, Program D. Membrane Concentrate Disposal: Practices and Regulation. VA, US: US Department of the Interior, Bureau of Reclamation, Technical Service; 2006

[36] Hoepner T, Lattemann S. Chemical impacts from seawater desalination plants — A case study of the northern Red Sea. Desalination. 2003;**152**(1):133-140. DOI: 10.1016/S0011-9164(02)01056-1

[37] Mohamed AMO, Maraqa M, Al HJ. Impact of land disposal of reject brine from desalination plants on soil and groundwater. Desalination. 2005;**182**(1):411-433. DOI: 10.1016/j.desal.2005.02.035

[38] Ariono D, Purwasasmita M, Wenten IG. Brine effluents: Characteristics, environmental impacts, and their handling. Journal of Engineering and Technological Sciences. 2016;**48**:367-387. DOI: 10.5614/j.eng.technol.sci.2016.48.4.1

[39] Jenkins S, Paduan J, Roberts P, Weis J. Management of Brine Discharges to Coastal Waters; Recommendations of a Science Advisory Panel. CA, USA: Southern California Coastal Water Research Project Costa. 2012

[40] Ahmed M, Anwar R. An assessment of the environmental impact of brine disposal in marine environment. International Journal of Modern Engineering Research. 2012;**24**:2756-2761

[41] Anders CR, Santos Fernandes C, da Silva DN, da Silva Gomes JW, de Souza Melo MR, de Souza BGA, et al. Environmental impacts of reject

brine disposal from desalination plants. Desalination and Water Treatment. 2020;**181**:17-26. DOI: 10.5004/dwt.2020.25055

[42] Pramanik BK, Shu L, Jegatheesan V. A review of the management and treatment of brine solutions. Environmental Science: Water Research & Technology. 2017;**3**(4):625-658. DOI: 10.1039/C6EW00339G

[43] Malfeito JJ, Díaz-Caneja J, Fariñas M, Fernández-Torrequemada Y, González-Correa JM, Carratalá-Giménez A, et al. Brine discharge from the Javea desalination plant. Desalination. 2005;**185**(1):87-94. DOI: 10.1016/j.desal.2005.05.010

[44] Jiménez-Arias D, Morales-Sierra S, Machado FJ, García-García A, Luis J, Valdés F, et al. Rejected brine recycling in hydroponic and thermo-solar evaporation systems for leisure and tourist facilities. Changing waste into raw material. Desalination. 2020;**496**:114443. DOI: 10.1016/j.desal.2020.114443

[45] Saf C, Gondet L, Villain-Gambier M, Belaqziz M, Trebouet D, Ouazzani N. Investigation of the agroecological applications of olive mill wastewater fractions from the ultrafiltration-nanofiltration process. Journal of Environmental Management. 2023;**333**:117467. DOI: 10.1016/j.jenvman.2023.117467

[46] Yuzer B, Aydin MI, Bulut FB, Palamutcu S, Emer U, Bekbolet M, et al. Reuse of the treated textile wastewater and membrane brine in the wet textile processes: Distorting effects on the cotton fabric. Desalination and Water Treatment. 2015;**56**(4):997-1009. DOI: 10.1080/19443994.2014.951001

[47] Management USEPAOoW, Branch USEPAOoWMMT. A Plain English Guide to the EPA Part 503 Biosolids Rule. Washington, DC: US Environmental Protection Agency, Office of Wastewater Management; 1994

[48] Banti DC, Tsangas M, Samaras P, Zorpas A. LCA of a membrane bioreactor compared to activated sludge system for municipal wastewater treatment. Membranes. 2020;**10**(12):1-15. DOI: 10.3390/membranes10120421

[49] Pontoni L, D'Alessandro G, d'Antonio G, Esposito G, Fabbricino M, Frunzo L, et al. Effect of anaerobic digestion on rheological parameters and dewaterability of aerobic sludges from MBR and conventional activated sludge plants. Chemical Engineering Transactions. 2015;**43**:2311-2316. DOI: 10.3303/CET1543386

[50] Fenu A, Roels J, Wambecq T, De Gussem K, Thoeye C, De Gueldre G, et al. Energy audit of a full scale MBR system. Desalination. 2010;**262**(1):121-128. DOI: 10.1016/j.desal.2010.05.057

[51] Massé A, Spérandio M, Cabassud C. Comparison of sludge characteristics and performance of a submerged membrane bioreactor and an activated sludge process at high solids retention time. Water Research. 2006;**40**(12):2405-2415. DOI: 10.1016/j.watres.2006.04.015

[52] Ding Z, Bourven I, Guibaud G, van Hullebusch ED, Panico A, Pirozzi F, et al. Role of extracellular polymeric substances (EPS) production in bioaggregation: Application to wastewater treatment. Applied Microbiology and Biotechnology. 2015;**99**(23):9883-9905. DOI: 10.1007/s00253-015-6964-8

[53] Pontoni L, Papirio S, D'Alessandro G, Caniani D, Gori R, Mannina G, et al. Dewaterability of CAS and MBR sludge: Effect of biological stability and EPS composition. Journal of Environmental Engineering.

2018;**144**(1):04017088. DOI: 10.1061/(ASCE)EE.1943-7870.0001299

[54] Semblante GU, Hai FI, Ngo HH, Guo W, You SJ, Price WE, et al. Sludge cycling between aerobic, anoxic and anaerobic regimes to reduce sludge production during wastewater treatment: Performance, mechanisms, and implications. Bioresource Technology. 2014;**155**:395-409. DOI: 10.1016/j.biortech.2014.01.029

[55] Lee LH, Wu TY, Shak KPY, Lim SL, Ng KY, Nguyen MN, et al. Sustainable approach to biotransform industrial sludge into organic fertilizer via vermicomposting: A mini-review. Journal of Chemical Technology & Biotechnology. 2018;**93**(4):925-935. DOI: 10.1002/jctb.5490

[56] Kurt M, Aksoy A, Sanin FD. Evaluation of solar sludge drying alternatives by costs and area requirements. Water Research. 2015;**82**:47-57. DOI: 10.1016/j.watres.2015.04.043

[57] Wang X, Xie Y, Qi X, Chen T, Zhang Y, Gao C, et al. A new mechanical cutting pretreatment approach towards the improvement of primary sludge fermentation and anaerobic digestion. Journal of environmental. Chemical Engineering. 2022;**10**(2):107163. DOI: 10.1016/j.jece.2022.107163

[58] Uddin MM, Wright MM. Anaerobic digestion fundamentals, challenges, and technological advances. Physical Sciences Reviews. 2022. DOI: 10.1515/psr-2021-0068

[59] Naserian ES, Cheraghi M, Lorestani B, Sobhanardakani S, Sadr MK. Qualitative investigation of sewage sludge composting: Effect of aerobic/anaerobic pretreatments. Arabian Journal of Geosciences. 2021;**14**(10):836. DOI: 10.1007/s12517-021-07232-x

[60] Houghton JI, Stephenson T. Effect of influent organic content on digested sludge extracellular polymer content and dewaterability. Water Research. 2002;**36**(14):3620-3628. DOI: 10.1016/S0043-1354(02)00055-6

[61] Asia I, Oladoja N, Bamuza-Pemu EE. Treatment of textile sludge using anaerobic technology. African Journal of Biotechnology. 2006;**5**(18):1678-1683

[62] Houghton JI, Quarmby J, Stephenson T. The impact of digestion on sludge Dewaterability. Process Safety and Environmental Protection. 2000;**78**(2):153-159. DOI: 10.1205/095758200530547

[63] Rorat A, Courtois P, Vandenbulcke F, Lemiere S. Sanitary and environmental aspects of sewage sludge management. In: Industrial and Municipal Sludge. Oxford, United Kingdom: Elsevier; 2019. pp. 155-180. DOI: 10.1016/B978-0-12-815907-1.00008-8

[64] Beshah DA, Tiruye GA, Mekonnen YS. Characterization and recycling of textile sludge for energy-efficient brick production in Ethiopia. Environmental Science and Pollution Research. 2021;**28**(13):16272-16281. DOI: 10.1007/s11356-020-11878-7

[65] Lissy P, Sreeja MJIJoM, e-ISSN CE. Utilization of sludge in manufacturing energy efficient bricks. 2014;**11**:1684-2278

[66] Patel H, Pandey S. Exploring the reuse potential of chemical sludge from textile wastewater treatment plants in India-a hazardous waste. American Journal of Environmental Sciences. 2009;**5**(1):106

[67] Jahagirdar SS, Shrihari S, Manu B. Utilization of textile mill sludge in burnt clay bricks. International Journal of Environmental Protection. 2013;**3**(5):6-13